Essays in Biochemistry

volume 30 1995

Essays in Biochemistry

Edited by D.K. Apps and K.F. Tipton

PORTLAND PRESS

Essays in Biochemistry is published by Portland Press Ltd on behalf of the Biochemical Society

Portland Press Ltd
59 Portland Place
London W1N 3AJ
U.K.

Although, at the time of going to press, the information contained in this publication is believed to be correct, neither the authors nor the publisher assume any responsibility for any errors or omissions herein contained. Opinions expressed in these Essays are those of the authors and are not necessarily held by the Biochemical Society, the editors or the publishers.

British Library Cataloguing-in-Publication Data
A catalogue record for this book is available from the British Library

ISBN 1-85578-018-6
ISSN 0071-1365

Typeset by Portland Press Ltd and printed in Great Britain
by Henry Ling Ltd, Dorchester

Contents

1 Contact sites and transport in mitochondria
Paul N. Moynagh

2 Tyrosine hydroxylase: human isoforms, structure and regulation in physiology and pathology
Toshiharu Nagatsu

3 **Antizyme-dependent degradation of ornithine decarboxylase**
Shin-ichi Hayashi

4 **The chloroplast genome**
Masahiro Sugiura

5 **Lectins — proteins with a sweet tooth: function in cell recognition**
Nathan Sharon and Halina Lis

9 **Thyrotropin-releasing hormone: basis and potential for its therapeutic use**
Julie A. Kelly

Authors

Paul N. Moynagh received his B.A. (Mod.) in 1988 and his Ph.D. in 1991 from Trinity College Dublin. He then spent one year as a CEC BRIDGE Programme postdoctoral fellow. In 1992 he was awarded an Irish Health Research Board postdoctoral fellowship. He was appointed a college lecturer at the Department of Pharmacology, University College Dublin in 1994. He is also a part-time tutor for the Open University.

Toshiharu (Toshi) Nagatsu obtained his M.D. in 1955 and his Ph.D. in 1960 from Nagoya University. He was Professor of Biochemistry at Aichi-Gakuin University School of Dentistry from 1966 to 1976, Professor of Cell Physiology at the Tokyo Institute of Technology from 1976 to 1985, Professor of Biochemistry at Nagoya University School of Medicine from 1984 to 1991. Since 1991, he has been Professor of Molecular Genetics and Neurochemistry at Fujita Health University School of Medicine. He has also worked at the National Institutes of Health, Bethesda, Maryland (1962–1964), at the University of Southern California, Los Angeles, California (1967–1968) and at the Roche Institute of Molecular Biology, Nutley, New Jersey (1972–1973). His research has centred on the biochemistry and molecular biology of biogenic amines, in particular catecholamines, and the related neurotransmitters.

Shin-ichi Hayashi obtained his M.D. in 1956 and his Ph.D. in 1962 from Osaka University Medical School. Between 1962 and 1964 he worked as a research fellow in the Department of Biological Chemistry at Harvard Medical School. He then held the position of Visiting Scientist at the National Institutes of Health, Bethesda, Maryland until 1966. In 1968 he was appointed Associate Professor in the Department of Nutrition and Physiological Chemistry at Osaka University Medical School. Since 1975 he has been Professor and Chairman of the Department of Nutrition at the Jikei University School of Medicine. His main research interests are in the field of metabolic regulation, especially the regulation of ornithine decarboxylase. Other studies include the regulation of bacterial glycerol metabolic system; dietary and hormonal regulation of tyrosine transaminase and serine synthesizing enzymes; metabolic disorder in obesity; and the mechanism of the hypercholesterolaemic effect of soy protein.

Masahiro Sugiura, Professor of Center for Gene Research, Nagoya University, was born in Okazaki in 1936 and graduated from Nagoya University with an M.Sc. in 1962. After 3 years training in the University of Illinois and the University of California at San Diego, he joined Hiroshima University, Kyoto University and then National Institute of Genetics where he started his work on the chloroplast genome. In 1982 he moved to Nagoya University.

Nathan Sharon and **Halina Lis** are associated with the Department of Membrane Research and Biophysics at the Weizmann Institute of Science, Rehovot, Israel. They have collaborated on studies of lectins, mainly from legumes, since the early 1960s. They have published extensively on the subject and together they have written close to 20 reviews, some very widely cited, as well as a book. Sharon has recently been awarded the prestigious Israel Prize for his research on lectins and complex carbohydrates, especially as related to cell recognition.

Alan Morgan has a B.Sc. in Applied Biology from Liverpool Polytechnic and a Ph.D in Physiology from the University of Liverpool. He is currently a lecturer in physiology at the University of Liverpool. For the past 8 years his research has focused on regulated exocytosis in adrenal chromaffin cells, covering areas such as Ca^{2+} signalling, signal transduction and the role of protein kinases. Current research projects concern the role of NSF, SNAPs and SNAREs in membrane traffic in general and in regulated exocytosis in particular.

Richard Ashley was born in Canada and moved to the U.K. He worked as a clinician for several years after graduating from St Thomas's Hospital Medical School, and obtained his Ph.D. in 1984 at the Institute of Psychiatry. After a period in the School of Biological Sciences at the University of Sussex, he took up a Wellcome Trust Research Training Fellowship in Mental Health at the National Heart and Lung Institute. In 1989 he joined the Department of Biochemistry at the University of Edinburgh Medical School, where he has been a senior lecturer since 1992.

John R. Murphy received his Ph.D. in Microbiology from the University of Connecticut School of Medicine in 1972. He was awarded a postgraduate fellowship and later received an Associate Professorship in the Department of Microbiology and Molecular Genetics at Harvard University. His research concerned the regulation, secretion and mechanism of diphtheria toxin action. In 1984 he was appointed Professor of Medicine at Boston University, and he is currently Chief of the Section of Biomolecular Medicine at Boston University Medical Center. His research work includes the development of hybrid toxin genes for the construction of eukaryotic cell receptor specific chimaeric toxins. **Éamonn B. Sweeney** graduated with a joint honours B.Sc. from St Patrick's College, Maynooth, in 1988. He developed an interest in the area of cancer treatment while working on the phenomenon of multidrug resistance at the Section of Medical Oncology in the Mater Hospital, Dublin. In 1995, he completed his Ph.D. at the Department of Biochemistry in Trinity College Dublin, and he is presently a postdoctoral research fellow at the Section of Biomolecular Medicine in Boston University Medical Center. He is currently developing cell-specific fusion proteins for the treatment of haematological malignancies.

Julie A. Kelly obtained her B.Sc. from Huddersfield Polytechnic in 1976. She was awarded an M.Sc. in 1977 and a Ph.D. in 1979 from the University of Manchester. On completion of her Ph.D. she was awarded a NATO Postdoctoral Fellowship to carry out research in the Department of Cell Biology at Baylor College of Medicine, Houston, Texas. Subsequently, she worked as a Research Scientist at the Center for Neurochemistry, New York. From 1983 to 1991 she was Assistant Professor in the Department of Chemistry at Manhattan College, New York; for a year during this period she also held the position of Visiting Lecturer in the Department of Biochemistry at the Royal College of Surgeons in Ireland, Dublin. From 1992 to 1994 she was appointed Senior Lecturer in Clinical Biochemistry in the Department of Biological Sciences at Manchester Metropolitan University. She currently holds the position of Research Fellow at Trinity College Dublin where she is investigating the functional role of thyrotropin-releasing hormone-degrading enzymes in the central nervous system.

Abbreviations

aa	amino acid
ACE	angiotensin-converting enzyme
ACh	acetylcholine
ALL	acute lymphoblastic leukaemia
AML	acute myelogenous leukaemia
ANC	adenine nucleotide carrier
AZ	antizyme
Ca/CaMPKII	Ca^{2+}/calmodulin-dependent protein kinase II
cADPR	cyclic ADP-ribose
CAT	chloramphenicol acetyltransferase
CDR	complementarity determining region
CHO	Chinese hamster ovary
CLL	chronic lymphocytic leukaemia
CK	creatine kinase
CNS	central nervous system
Cr	creatine
CRD	carbohydrate-recognition domain
CRE	cyclic AMP-response element
CREB	CRE-binding protein
ctDNA	chloroplast DNA
cytochrome P-450 scc	cytochrome P-450 side-chain cleavage enzyme
DFMO	difluoromethylornithine
DHPR	dihydropyridine receptor
DRD	dopa-responsive dystonia
DT	diphtheria toxin
EC-coupling	excitation–contraction-coupling
EF-2	elongation factor 2
EGF	epidermal growth factor
ER	endoplasmic reticulum
ERK	extracellular signal-regulated kinase
F3	Fluo-3
FKBP	FK506-binding protein
GABA	γ-aminobutyric acid
GM-CSF	granulocyte/macrophage colony stimulating factor
gp	glycoprotein
HAMA	human anti-mouse antibody
HPD	hereditary progressive dystonia
hsp	heat-shock protein

HTC	hepatoma tissue culture
IL-2R	interleukin-2 receptor
IL	interleukin
i.m.	intramuscular
Ins(1,4,5)P_3	inositol 1,4,5-trisphosphate
InsP_3R	inositol 1,4,5-trisphosphate receptor
i.p.	intraperitoneal
IPTG	isopropyl β-D-thiogalactoside
IR	inverted repeat
ISP	import site protein
IT	immunotoxin
i.v.	intravenous
LDCV	large dense-core vesicle
LTP	long-term potentiation
mAb	monoclonal antibody
MAP kinase	mitogen-activated protein kinase
MAPKAP kinase	MAP kinase-activated kinase
MBR	mitochondrial benzodiazepine receptor
MIM	mitochondrial inner membrane protein
MND	motor neuron disease
MOM	mitochondrial outer membrane protein
MSH	melanocyte-stimulating hormone
NeuAc	N-acetylneuraminic acid
NGF	nerve growth factor
NMDA	N-methyl-D-aspartate
NSF	N-ethylmaleimide-sensitive fusion protein
ODC	ornithine decarboxylase
opamp	operational amplifier
Orn	ornithine
PAP	pyroglutamyl aminopeptidase
pAb	polyclonal antibody
PCr	phosphocreatine
PCR	polymerase chain reaction
PE	*Pseudomonas* exotoxin A
PKA	protein kinase A
PKC	protein kinase C
PLGA	copoly(D,L-lactic/glycolic acid)
Put	putrescine
RT-PCR	reverse transcription-polymerase chain reaction
RyR	ryanodine receptor
s.c.	subcutaneous
SiaLe[a]	sialyl-Lewis[a]

SiaLe^x	sialyl-Lewis^x
SNAP	soluble NSF-attachment protein
SNAP-25	synaptosomal-associated protein, 25 kDa
SNARE	SNAP receptor
Spd	spermidine
Spm	spermine
SR	sarcoplasmic reticulum
SV	synaptic vesicle
TH	tyrosine hydroxylase
TRH	thyrotropin-releasing hormone
TSH	thyroid-stimulating hormone
Ub	ubiquitin
VAMP	vesicle-associated membrane protein
VDAC	voltage-dependent anion channel
VIP	vasoactive intestinal peptide
WGA	wheat germ agglutinin

I

Contact sites and transport in mitochondria

Paul N. Moynagh

Department of Pharmacology, University College Dublin, Blackrock, Dublin 4, Republic of Ireland

Introduction

The mitochondrion is the cellular organelle responsible for oxidative phosphorylation and thus aerobic metabolism in eukaryotes. This fundamental process relies on the crucial involvement of ion gradients across the innermost of the two mitochondrial membranes. Therefore, the mitochondrial inner membrane imposes a very strict control on permeability to ions by employment of highly selective carriers. In contrast, the mitochondrial outer membrane is often considered leaky and acts merely as a crude sieve between the energy-transducing inner membrane and the cytosol. However, recent research has suggested that components of the outer membrane, especially those regions which come into close contact with the inner membrane, may have more subtle roles with respect to overall mitochondrial function. This essay reviews such recent developments and considers the implications that these 'contact sites' may play an integral part not only in regulating mitochondrial physiology but also in the biogenesis of the organelle.

Composition of contact sites

Contact sites, describing specific regions of physical contact between mitochondrial inner and outer membranes, were first reported by Hackenbrock[1] in 1968 when analysing isolated liver mitochondria by high-resolution electron microscopy. Such contact sites explain deflections in the fracture plane of freeze-fractured mitochondria, characterized by frequent jumping of the fracture between the adjacent membranes[2]. Indeed, the frequency of deflections has been used to quantify the frequency of contact sites[3,4]. Such studies have demonstrated the dynamic nature of contact sites. Thus isolated mitochondria that are actively engaged in the phosphorylation of ADP possess more contact sites than resting mitochondria without ADP. This is explained by contact site induction in response to ADP. In addition, contact frequency is regulated in intact cells as has been shown in cultured hepatocytes where contacts were induced by adrenaline but decreased in frequency in the presence of glucagon[5]. The critical dependence of contact site formation on the functional status of the mitochondrion has prompted much research into the polypeptide composition of contact sites. This has led to the identification of a number of components, some of which are integral proteins of the inner or outer membranes, while others specifically associate with contacts (Figure 1).

The ADP/ATP translocator, an integral inner membrane protein, was probed for its distribution in submitochondrial fractions enriched in contacts or in inner membrane devoid of contact regions. The translocator was found in both fractions; however, when compared with the distribution of cytochrome oxidase, the translocator was present at a 3-fold higher concentration in the contact sites[6]. The presence of the translocator in regions of the inner membrane that are not involved in contact site formation is hardly surprising given the dynamic nature of contacts. This is also true of a channel-forming protein, voltage-dependent anion channel (VDAC; also known as mitochond-

Figure 1. Schematic representation of contact site region in mitochondria
Abbreviations used: ANC, adenine nucleotide carrier; CK, creatine kinase; MBR, mitochondrial benzodiazepine receptor; VDAC, voltage-dependent anion channel.

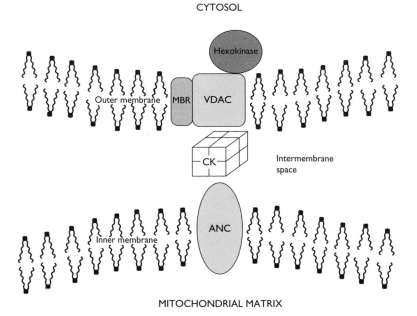

rial porin), which is an integral outer membrane protein constituting part of the contact site structure and is also present in non-contact regions of the outer membrane[7]. VDAC, a 30 kDa polypeptide, forms a voltage-gated ion channel; however, unlike other such channels which are formed by bundles of α-helices, VDAC forms its channel via a cylindrical β-barrel[8]. Thus hydropathy plots do not show predominantly hydrophobic 20-residue runs commonly considered to be indicators of transmembrane helices. Instead, alternating polar and non-polar amino acids are conducive to β-sheet formation with distinct polar and non-polar sides. The polar side would be predicted to line the lumen of the channel, while the non-polar side would lie buried in a hydrophobic membrane environment. The properties of the channel formed by VDAC dictate the permeability of the outer membrane to small inorganic ions. Thus the channel, which has a reported pore radius of 2.5 nm, is extremely permeable to small inorganic ions, demonstrating a conductance 35 times greater than that exhibited by a rat brain sodium channel. The channel is not ion specific but is selectively more permeable to anions than cations. It is voltage gated which is reflected in its switch to lower-conducting substates (300 pS; note that 1 pS $\equiv 6.3 \times 10^6$ ions conducted per s per V) when exposed to transmembrane potentials as low as 20 mV. This voltage-induced 'closed' state also demonstrates opposite ion selectivity, with a greater preference for cations. The closed state of the channel is also induced by a synthetic anionic polymer[9] and by an endogenous mitochondrial modulator[10]. Interestingly, the former has been demonstrated to inhibit adenine nucleotide transport through the channel[11], thus precluding access of the nucleotides to the ADP/ATP translocator in the inner membrane. The closing of the channel by either application of a transmembrane potential or binding of macromolecular ligands may, therefore, act as a key regulatory device in the control of mitochondrial function[12].

Contact sites contain a number of kinases. Klug and co-workers[4] first noted that the frequency of contacts correlated with hexokinase activity. Subsequent immunogold electron microscopy localized hexokinase to the surface of the outer membrane of mitochondria but only in regions where contacts were apparent[13]. However, previous studies had reported that VDAC was the hexokinase-binding protein in mitochondria[14,15]. Thus VDAC is randomly distributed in the outer membrane, whereas hexokinase is restricted to contact regions. The conclusion, therefore, is that VDAC in contact sites displays certain features necessary for hexokinase binding which are not apparent in VDAC outside contacts. Two additional kinases, nucleoside-diphosphate kinase (EC 2.7.4.6) and creatine kinase (EC 2.7.3.2), constitute part of the contact site structure but these are located between the two membranes[7,16]. Creatine kinase seems especially suited to link the inner and outer membranes together (reviewed by Wallimann et al.[17]). This octameric species consists of four identical dimers forming a cube-like molecule with a central channel running through the middle. The top and bottom faces of the octamer confer on it the ability to interact simultaneously with components of the inner and outer membranes and thus facilitate a closer association of the two membranes (Figure 1).

A recent study by McEnery et al.[18] described the isolation of the mitochondrial benzodiazepine receptor from rat kidney with the concomitant purification of the ADP/ATP translocator and VDAC. The mitochondrial benzodiazepine receptor is distinct from the neuronal plasma membrane receptor, which is associated with the GABA$_A$ receptor (GABA, γ-aminobutyric acid) and through which the pharmacological effects of benzodiazepines are mediated. Instead, the mitochondrial receptor is present in many peripheral tissues as well as in the central nervous system[19]. It has been localized to the mitochondrial outer membrane[20,21] where it binds certain benzodiazepines[19] and isoquinoline carboxamides[22] with nanomolar affinities. The receptor demonstrates a selective requirement for phosphatidylserine to exhibit such binding capabilities[23]. Many different roles have been proposed for ligands of the receptor, including inhibition of respiratory control[24] and stimulation of steroidogenesis[25]. However, the role of the receptor in mediating these effects remains to be determined. The recent purification of the receptor in close association with the ADP/ATP translocator and VDAC, coupled with the absolute requirement of these two components for receptor activity, suggests a contact site location for the receptor which may be an important aspect for its function.

A model of complexity emerges from the above explanation for the structure of contact sites. Such regions contain the ADP/ATP translocator and VDAC in their inner and outer membrane domains, respectively. The membranes may be 'glued' together via creatine kinase and nucleoside-diphosphate kinase, with hexokinase and a benzodiazepine receptor located on the surface of the outer membrane. Such a multi-component complex may play a critical role in the functioning of mitochondria.

Contact sites and mitochondrial function

Energy metabolism

The higher frequency of contact sites in mitochondria engaged in the phosphorylation of ADP relative to resting mitochondria without ADP indicates that contact sites may be involved in regulating energy metabolism in mitochondria. Indeed, contact site induction correlates with ADP level, the latter being a key component in modulating respiration. Thus increased ATP usage in the cytoplasm results in an increase in ADP concentration and this increase in cytosolic ADP will stimulate ADP uptake in exchange for ATP export since the flux through the ADP/ATP translocator is sensitive to the mitochondrial and cytosolic free ADP and ATP levels. Efficient exchange requires that ATP must diffuse from mitochondria to regions of utilization in the cell, whereas ADP must diffuse in the opposite direction. Inefficiencies in these processes

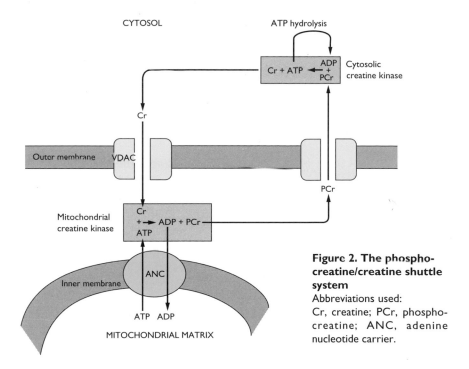

Figure 2. The phospho-
creatine/creatine shuttle
system
Abbreviations used:
Cr, creatine; PCr, phosphocreatine; ANC, adenine
nucleotide carrier.

may result in the ADP/ATP translocator limiting respiration rate. A model involving the generation of contact sites has been proposed which avoids potential diffusion problems and this is of relevance to excitable tissues, such as muscle[17,26] (Figure 2). A phosphocreatine/creatine shuttle system operates to form an 'energy transport' system which depends on the interaction of cytoplasmic and mitochondrial forms of creatine kinase. The mitochondrial form consists of an octamer which has been shown to induce the formation of contact sites between purified mitochondrial inner and outer membranes[27], probably via its two symmetrical hydrophobic sides. It has thus been proposed that the mitochondrial creatine kinase can simultaneously associate with the ADP/ATP translocator in the inner membrane and VDAC in the outer membrane. The creatine kinase, in this location, has optimal access to mito-chondrially generated ATP and uses it to phosphorylate creatine. Concomitantly, the ADP/ATP translocator becomes saturated with ADP and is primed for another ATP/ADP antiport reaction. Such a scenario has been proposed on the basis of experiments demonstrating that ADP produced by mitochondrial creatine kinase does not readily equilibrate with the extramito-chondrial ADP/ATP pool[5]. The other product of the reaction catalysed by mitochondrial creatine kinase, phosphocreatine, may diffuse to the cytosolic isoenzymes of creatine kinase which can use the phosphocreatine to phospho-rylate ADP generated locally; the creatine can then diffuse back to the mitochondria. Thus in this system phosphocreatine and creatine are acting as the transport molecules in the cell and are probably better suited for this purpose than ATP and ADP, since they are smaller compounds with a low negative charge compared with the adenine nucleotides. Indeed, this proposal has been confirmed recently by non-invasive ^{31}P-pulsed gradient NMR measurements *in vivo*[28]. These studies indicated that the mean diffusion lengths of phosphocreatine and creatine in the cell (57 µm and 37 µm, respectively) are significantly higher than those of ADP and ATP (1.8 µm and 22 µm, respectively).

The translocation of phosphocreatine and creatine through the mitochondrial outer membrane is suggested to be restricted to VDAC. As stated previously, VDAC can assume a high-conductance, anion-selective state or a voltage-induced 'closed' state demonstrating cation selectivity. It has been proposed that the latter is induced at mitochondrial contact sites by capacitative coupling between the inner and outer membranes, whereby the inner membrane potential influences the polarity across the outer membrane[5]. In this model, the transfer of positive charges through the inner membrane (e.g. proton transfer) would result in the generation of a field across the outer membrane with polarity outside negative. Such coupling requires the two membranes to be closely apposed, as is the case in contact sites, and the resulting field across the outer membrane would impose a cation selectivity on VDAC in these regions. In contrast, VDAC beyond the contact sites would remain anion selective and unregulated since the separation distance between the two membranes is too large to allow capacitative coupling. Thus creatine would enter at cation-selective pores at the contact regions, whereas phosphocreatine would exit through the anion-selective pores just beyond the contacts (Figure 2). The inaccessibility of VDAC to phosphocreatine at contact sites would mask the mitochondrial creatine kinase at this site from high concentrations of phosphocreatine. This would allow creatine kinase at contact sites to produce creatine phosphate from creatine and ATP far beyond the equilibrium that would be imposed in the absence of the masking effect. Evidence for this emerges from experiments which demonstrate that exogenously added phosphocreatine has no inhibitory effect on the reaction of creatine kinase in intact brain mitochondria. It is thus hypothesized that the ADP/ATP translocator, the intermembrane mitochondrial creatine kinase and VDAC form an extended, highly organized, multi-enzyme energy-transport channel through the inner and outer membranes at contact sites.

Contact sites are associated with other kinases which may well be of significance in additional energy coupling reactions. Therefore, hexokinase is both bound and activated by VDAC but only when the latter is present in contact sites. Hexokinase in this location is suitably placed to benefit from mitochondrially produced ATP emerging from the ADP/ATP translocator and, as a consequence, a very efficient energy transfer occurs. This may be a very important aspect of glucose uptake into cells especially in the brain[5]. Increased ATP utilization causes an increase in ADP levels which induces the formation of contact sites. This results in the recruitment and activation of hexokinase which ultimately leads to increased cellular uptake of glucose. Thus the importance of contact sites may extend to overall cell metabolism.

Interestingly, ligands for another component of contact sites, the mitochondrial benzodiazepine receptor, have been shown to inhibit respiratory control by increasing the rate of substrate oxidation (state 4) and decreasing the rate of oxidative phosphorylation (state 3)[24]. While the mechanism remains to be determined, it is possible that the benzodiazepine receptor in contact sites may be involved in limiting and/or facilitating exchange of substrates and products between mitochondria and the cytosol. The above report, in conjunction with research on mitochondrial creatine kinase and hexokinase, would argue strongly in favour of contact sites and their components performing key roles in regulating energy metabolism in cells.

Intramitochondrial cholesterol transfer and steroidogenesis[25]

The first step in steroidogenesis involves the conversion of cholesterol to pregnenolone by cytochrome P-450 side-chain cleavage (P-450 scc) enzyme in the mitochondrial inner membrane. Pregnenolone is then transferred to the smooth endoplasmic reticulum where various enzymic reactions result in many steroid products. The rate limiting step in the overall process is the transport of cholesterol from extramitochondrial stores to the mitochondria, and its subsequent distribution within the mitochondria to cytochrome P-450 scc enzyme. The finding that the mitochondrial benzodiazepine receptor is most abundant in steroidogenic cells, coupled with its location on the mitochondrial outer membrane, suggested a possible role for the receptor in steroid biosynthesis. Indeed, it was subsequently shown in steroidogenic cells and isolated mitochondria that nine different ligands for the benzodiazepine receptor, with affinities spanning four orders of magnitude, stimulated steroid production with potencies which correlated with their affinities for the receptor. The step of the steroid biosynthetic pathway modulated by the benzodiazepine receptor was investigated and it was found that the binding of ligands to the receptor caused an increase in translocation of cholesterol from the mitochondrial outer to inner membrane. It thus appears that the mitochondrial benzodiazepine receptor modulates steroid production at the step of intramitochondrial cholesterol transport by regulating substrate supply to cytochrome P-450 scc, the first enzyme in the steroid biosynthetic pathway[25].

Surprisingly, flunitrazepam, a benzodiazepine that binds to the mitochondrial benzodiazepine receptor, inhibits hormone-stimulated steroid production in steroidogenic cells by reducing cholesterol transport from the mitochondrial outer to inner membranes. Such an inhibitory effect was precluded upon addition of higher affinity ligands for the benzodiazepine receptor. Interestingly, flunitrazepam has been shown to photolabel a 30 kDa protein in mitochondria and this has recently been identified to be VDAC[18]. As stated previously, VDAC complexes with the mitochondrial benzodiazepine receptor in the outer membranes. Since it has been demonstrated that VDAC contains associated cholesterol (five molecules per polypeptide chain in the case of mitochondria from bovine heart), it is tempting to propose that the binding of high-affinity ligands to the benzodiazepine receptor may effect the release of cholesterol from VDAC and thus make it available for transfer to the mitochondrial inner membrane. If this is the case it is likely that such a process occurs at contact sites. This is based on the finding that the mitochondrial benzodiazepine receptor was recently purified in a form complexed not only to VDAC but also to the ADP/ATP translocator and, in addition, the latter two components were essential for receptor activity[18]. Thus contact sites may well represent highly organized complexes which bring the outer and inner membranes of the mitochondria into close proximity and facilitate cholesterol transfer between the membranes, the key regulatory step in steroidogenesis

Phospholipid biosynthesis

Much research has recently focused on interorganelle movement of lipids within cells. A well-studied model is phosphatidylserine metabolism (Figure 3). The synthesis of phosphatidylserine from serine is catalysed by phosphatidylserine synthase (EC 2.7.8.8) which is found predominantly in the endoplasmic reticulum of liver. Once formed the phosphatidylserine must

Figure 3. Phosphatidylserine metabolism and associated phospholipid biosynthesis
Abbreviations used: PC, phosphatidylcholine; PE, phosphatidylethanolamine; PS, phosphatidylserine.

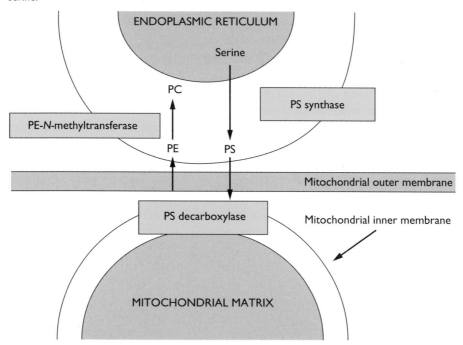

be transported to the mitochondrial inner membrane where it is acted on by phosphatidylserine decarboxylase (EC 4.1.1.65) to form phosphatidylethanolamine. In liver the subsequent methylation of phosphatidylethanolamine apparently provides up to 30% of the cellular phosphatidylcholine. However, phosphatidylethanolamine-*N*-methyltransferase (EC 2.1.1.17), the enzyme which catalyses the three successive methylations of phosphatidylethanolamine, is located in the endoplasmic reticulum. Thus the methylation of phosphatidylserine-derived phosphatidylethanolamine requires the transport of phosphatidylethanolamine from the mitochondrial inner membrane to the endoplasmic reticulum. It is clear that in this system of concerted phospholipid biosynthesis, interorganellar trafficking of lipids serves an integral function.

It was shown using reconstitution experiments with isolated organelles derived from liver that phosphatidylserine translocation between the endoplasmic reticulum and the mitochondria can occur in the absence of soluble factors[29]. This lent support to the idea of a collision-based mechanism in which the two organelles come in close contact with each other and thus facilitate phospholipid transfer. Further evidence for such direct organelle interaction emerged from the isolation from rat liver of a membrane fraction which was associated with mitochondria and had high specific activity for several phospholipid biosynthetic enzymes that are usually designated as being located in the endoplasmic reticulum[30]. This membrane fraction with its associated mitochondria was capable of the concerted synthesis of phosphatidylserine, phosphatidylethanolamine and phosphatidylcholine from serine. An interesting finding was that the inhibition of phosphatidylserine decarboxylase led to a preferential accumulation of the translocated phosphatidylserine in contact-site-enriched fractions, providing strong evidence that the intramitochondrial translocation of microsomal phosphatidylserine to the inner membrane occurs at contact sites[31]. It was also demonstrated that newly synthesized phosphatidylethanolamine in the inner membrane is also transported to the mitochondrial outer membrane via contact sites. In addition, recent studies have shown that the endoplasmic reticulum interacts with the mitochondria at contact site regions[32]. Taken together, these results support a model in which contact sites in mitochondria provide docking regions for endoplasmic reticulum, thereby facilitating efficient transfer of newly synthesized phosphatidylserine from the endoplasmic reticulum to the mitochondria (Figure 3). The contacts also participate in the intramitochondrial translocation of phosphatidylserine from the outer membrane to phosphatidylserine decarboxylase in the inner membrane and the subsequent export of newly formed phosphatidylethanolamine to the mitochondrial surface, where the endoplasmic reticulum is anchored and ready to catalyse the methylation of phosphatidylethanolamine to phosphatidylcholine. Thus it appears that contact sites in this context not only serve in internal affairs of the mitochondria but also allow for functional interaction of mitochondria with other cellular organelles.

It has been shown that the stability of the mitochondrial benzodiazepine receptor with respect to ligand binding activity is highly dependent on an appropriate lipid environment[23]. Interestingly, of all the lipids tested, phosphatidylserine proved the most efficacious in maintaining receptor activity subsequent to detergent solubilization. This finding, in conjunction with the putative contact site location for the receptor, suggests that this benzodiazepine receptor may be intimately involved in phosphatidylserine metabolism. Such circumstantial evidence adds further to the proposal that contact sites and their components are crucial in the phospholipid biosynthetic pathways associated with phosphatidylserine metabolism.

Contact sites involved with mitochondrial protein import

The targeting of proteins to mitochondria includes both intramitochondrial sorting of proteins encoded by the organellar genome, and import and subsequent sorting of nuclear-encoded precursor proteins. The mitochondrial genome encodes only a few proteins and these are all directed to the mitochondrial inner membrane. The majority of mitochondrial proteins are nuclear encoded and synthesized as precursors in cytoplasm. The precursors bind to outer membrane receptor proteins and are then translocated across the outer and inner membranes into the mitochondrial matrix[33,34]. Translocation appears to occur at sites where the two membranes are closely apposed. Recently, much effort has been concentrated on elucidating the identity of the components constituting these protein translocation 'contact sites'. A complex picture emerges, but the suggestion is that these contact sites are distinct from those previously described in the context of energy metabolism and intramitochondrial lipid transfer.

The process of protein import from the cytosol to the mitochondrial matrix is outlined in Figure 4. Mitochondrially targeted proteins are synthesized on cytosolic ribosomes as preproteins which possess positively charged signal sequences (presequences) at their N-termini. These preproteins are guided by cytosolic chaperones and mitochondrial outer membrane receptors to the outer membrane translocation complex which is present in contact site regions. This hetero-oligomeric complex spans the outer membrane and may form a translocation pore. Some of the polypeptide subunits of the complex have been identified in *Neurospora crassa*. These include mitochondrial outer membrane proteins MOM38, MOM7, MOM8, MOM22 and MOM30, where the numbers represent the molecular mass of each subunit. Equivalent subunits have been identified in yeast[35]; in addition, ISP6, a small hydrophobic outer membrane protein in yeast, has been shown to interact with the homologue of MOM38. At present, it is unclear how all of these subunits interact to effect the transfer of preproteins across the mitochondrial outer membrane. It is

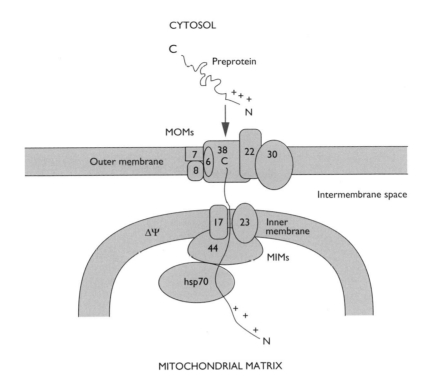

Figure 4. Pathway for import of proteins into mitochondria
Abbreviations used: MIM, mitochondrial inner membrane protein; MOM, mitochondrial outer membrane protein.

presumed that they form a translocation channel which serves to receive incoming preproteins from the outer membrane receptors and subsequently to guide preproteins to an adjacent mitochondrial inner membrane import complex.

Recently the first components of the mitochondrial inner membrane import machinery have been described in yeast[34]. Thus mitochondrial inner membrane proteins MIM17 and MIM23 behave as integral proteins of the mitochondrial inner membrane and may form the preprotein translocation channel of the inner membrane. MIM44 [import site protein45 (ISP45) in yeast] is the third essential component of the inner membrane import machinery. It seems to be a peripheral membrane protein on the mitochondrial matrix side but is firmly associated with the inner membrane by interaction with integral membrane proteins (possibly MIM17 or MIM23). MIM44 (ISP45) can be specifically cross-linked to a preprotein in transit across the mitochondrial membranes, suggesting that MIM44 may function in preprotein binding during import. Heat-shock protein hsp70 in the mitochondrial matrix also binds to preprotein in transit. Indeed, since a significant proportion of this hsp70 is reversibly associated with MIM44, it has been proposed that the two proteins function in close cooperation to drive protein import through the inner membrane into the mitochondrial matrix. Whereas the transfer of protein across the outer membrane is energy independent, the movement of polypeptide chains across the mitochondrial inner membrane requires the input of two sources of energy. A membrane potential across the inner membrane (negative inside) is required for translocation of the preprotein presequence. As stated previously, the presequence is positively charged and thus it may be that presequence translocation occurs by an electrophoretic mechanism. The translocation of the remainder of the preprotein is independent of membrane potential but requires ATP from the mitochondrial matrix. This ATP requirement seems to be due to an ATP-dependent action of mitochondrial hsp70.

The current model favours the idea that protein-translocation contact sites constitute the linking of protein import systems of the mitochondrial outer and inner membranes. The transport machineries of both membranes are not permanently connected but can be transiently linked by translocating polypeptide chains which span both mitochondrial membranes simultaneously. Contact sites are thus formed in the process. The dynamic nature of these sites resembles that of the contact sites involved in energy metabolism and intra-mitochondrial lipid transfer. However, this may be the limit of their relationship since, owing to their quite different polypeptide compositions, they seem to constitute distinct types of contact site.

Conclusions

Contact sites are not merely morphological features describing regions where mitochondrial outer and inner membranes are brought into close proximity; they play key roles in energy metabolism and lipid trafficking within mitochondria and, in the context of these processes, they also facilitate the functional interfacing of mitochondria with other components of the cell. This highlights the global importance of contact sites with respect to overall cell activity. In addition, the rapidly expanding field of mitochondrial protein import has implicated contact sites to be integral components in the biogenesis of the organelle. Thus contact sites may be considered as structures which execute roles of fundamental importance to the generation and subsequent activity of mitochondria.

My research is supported by the Irish Health Research Board, Forbairt and the CEC BIOTECH program (CT 93-0224). I would like to thank Professor Clive Williams for helpful discussion and Dr Sinead Boyce for typing the manuscript.

References

1.   Hackenbrock, C.R. (1968) Chemical and physical fixation of isolated mitochondria in low-energy and high-energy states. Proc. Natl. Acad. Sci. U.S.A. **61**, 598–605

2.   van Venetie, R. & Verkleij, A.J. (1982) Possible role of non-lipids in the structure of mitochondria: a freeze-fracture electron microscopy study. *Biochim. Biophys. Acta* **692**, 397–405

3.   Knoll, G. & Brdiczka, D. (1983) Changes in freeze-fractured mitochondrial membranes correlated to their energetic state. *Biochim. Biophys. Acta* **733**, 102–110

4.   Klug, G.A., Krause, J., Ostlund, A., Knoll, G. & Brdiczka, D. (1984) Alterations in liver mitochondrial function as a result of fasting and exhaustive exercise. *Biochim. Biophys. Acta* **764**, 272–282

5.   Brdiczka, D. (1991) Contact sites between mitochondrial envelope membranes. Structure and function in energy- and protein-transfer. *Biochim. Biophys. Acta* **1071**, 291–312

6.   Kottke, M., Adams, V., Wallimann, T., Nalam, V.K. & Brdiczka, D. (1991) Location and regulation of octameric mitochondrial creatine kinase in the contact sites. *Biochim. Biophys. Acta* **1061**, 215–225

7.   Adams, V., Bosch, W., Schlegel, J., Wallimann, T. & Brdiczka, D. (1989) Further characterization of contact sites from mitochondria of different tissues: topology of peripheral kinases. *Biochim. Biophys. Acta* **981**, 213–225

8.   Mannella, C.A. (1990) Structural analysis of mitochondrial pores. *Experientia* **46**, 137–145

9.   Colombini, M., Yeung, C.H., Tung, J. & Konig, T. (1987) The mitochondrial outer membrane channel, VDAC, is regulated by a synthetic polyanion. *Biochim. Biophys. Acta* **905**, 279–286

10. Holden, M.J. & Colombini, M. (1988) The mitochondrial outer membrane channel, VDAC, is modulated by a soluble protein. *FEBS Lett.* **241**, 105–109

11. Benz, R., Wojtczak, L., Bosch, W. & Brdiczka, D. (1988) Inhibition of adenine nucleotide transport through the mitochondrial porin by a synthetic polyanion. *FEBS Lett.* **231**, 5–80

12   Liu, M.Y. & Colombini, M. (1992) Regulation of mitochondrial respiration by controlling the permeability of the outer membrane through the mitochondrial channel, VDAC. *Biochim. Biophys. Acta* **1098**, 255–260

13. Kottke, M., Adam, V., Riesinger, I. *et al.* (1988) Mitochondrial boundary membrane contact sites in brain: points of hexokinase and creatine kinase location, and control of Ca^{++} transport. *Biochim. Biophys. Acta* **935**, 87–102

14. Linden, M., Gellerfors, P. & Nelson, B.D. (1982) Pore protein and the hexokinase-binding protein from the outer membrane of rat liver mitochondria are identical. *FEBS Lett.* **141**, 189–192

15. Fiek, C., Benz, R., Roos, N. & Brdiczka, D. (1982) Evidence for identity between the hexokinase-binding protein and the mitochondrial porin in the outer membrane of rat liver mitochondria. *Biochim. Biophys. Acta* **688**, 429–440

16. Biermans, W., Bernaert, I., De Bie, M., Nijs, B. & Jacob, W. (1989) Ultrastructural localisation of creatine kinase activity in the contact sites between inner and outer mitochondrial membranes of rat myocardium. *Biochim. Biophys. Acta* **974**, 74–80

17. Wallimann, T., Wyss, M., Brdiczka, D., Nicolay, K. & Eppenberger, H.M. (1992) Intracellular compartmentation, structure and function of creatine kinase Isoenzymes in tissues with high and fluctuating energy demands: the 'phosphocreatine circuit' for cellular energy homeostasis. *Biochem. J.* **281**, 21–40

18. McEnery, M.W., Snowman, A.M., Trifiletti, R.R. & Snyder, S.H. (1992) Isolation of the mitochondrial benzodiazepine receptor: association with the voltage-dependent anion channel and the adenine nucleotide carrier. *Proc. Natl. Acad. Sci. U.S.A.* **89**, 3170–3174

19. Braestrup, C. & Squires, R.F. (1977) Specific benzodiazepine receptors in rat brain characterised by high-affinity [^{3}H]diazepam binding. *Proc. Natl. Sci. Acad. U.S.A.* **74**, 3805–3809

20. Anholt, R.R.H., Pederson, P.L., De Souza, E.B. & Snyder, S.H. (1986) The peripheral-type benzodiazepine receptor: localization to the mitochondrial outer membrane. *J. Biol. Chem.* **261**, 576–583

21. O'Beirne, G.B. & Williams, D.C. (1988) The subcellular location in rat kidney of the peripheral benzodiazepine acceptor. *Eur. J. Biochem.* **175**, 413–421

22. Le Fur, G., Perrier, M.L., Vaucher, N. *et al.* (1983) Peripheral benzodiazepine binding sites: effects of PK 11195, 1-(2-chlorophenyl)-N-(1-methylpropyl)-3-isoquinolinecarboxamide. I. *in vitro* studies. *Life Sci.* **32**, 1839–1847

23. Moynagh, P.N. & Williams, D.C. (1992) Stabilization of the peripheral-type benzodiazepine acceptor by specific phospholipids. *Biochem. Pharmacol.* **43**, 1939–1945

24. Hirsch, J.D., Beyer, C.F., Malkowitz, L., Beer, B. & Blume, A.J. (1989) Mitochondrial benzo-diazepine receptors mediate inhibition of mitochondrial respiratory control. *Mol. Pharmacol.* **35**, 157–163

25. Papadopoulos, V. (1993) Peripheral-type benzodiazepine/diazepam binding inhibitor receptor: biological role in steroidogenic cell function. *Endocr. Rev.* **14**, 222–240

26. Brdiczka, D. (1994) Function of the outer mitochondrial compartment in regulation of energy metabolism. *Biochim. Biophys. Acta* **1187**, 264–269

27. Rojo, M., Hovius, R., Demel, R.A., Nicolay, K. & Wallimann, T. (1991) Mitochondrial creatine kinase mediates contact formation between mitochondrial membranes. *J. Biol. Chem.* **266**, 20290–20295

28. Yoshizaki, K., Watari, H. & Radda, G.K. (1990) Role of phosphocreatine in energy transport in skeletal muscle of bullfrog studied by ^{31}P-n.m.r. *Biochim. Biophys. Acta.* **1051**, 44–150

29. Voelker, D.R. (1989) Reconstitution of phosphatidylserine import into rat liver mitochondria. *J. Biol. Chem.* **264**, 8019–8025

30. Vance, J.E. (1990) Phospholipid synthesis in a membrane fraction associated with mitochondria. *J. Biol. Chem.* **265**, 7248–7256

31. Ardail, D., Lerme, F. & Louisot, P. (1991) Involvement of contact sites in phosphatidylserine import into liver mitochondria. *J. Biol. Chem.* **266**, 7978–7981

32. Ardail, D., Gasnier, F., Lerme, F., Simonot, C., Louisot, P. & Gateau-Roesch, O. (1993) Involvement of mitochondrial contact sites in the subcellular compartmentalization of phospho-lipid biosynthetic enzymes. *J. Biol. Chem.* **268**, 25985–25992

33. Hannavy, K., Rospert, S. & Schatz, G. (1993) Protein import into mitochondria: a paradigm for the translocation of polypeptides across membranes. *Curr. Opin. Cell Biol.* **5**, 694–700

34. Pfanner, N., Craig, E.A. & Meijer, M. (1994) The protein import machinery of the mitochondrial inner membrane. *Trends Biochem. Sci.* **19**, 368–372

35. Moczko, M., Dietmeier, K., Söllner, T. *et al.* (1992) Identification of the mitochondrial receptor complex in *Saccharomyces cerevisiae*. *FEBS Lett.* **310**, 265–268

2

Tyrosine hydroxylase: human isoforms, structure and regulation in physiology and pathology

Toshiharu Nagatsu

Institute for Comprehensive Medical Science, School of Medicine, Fujita Health University, Toyoake, Aichi 470-1, Japan

Introduction

Tyrosine 3-hydroxylase (TH; EC 1.14.16.2) catalyses the first step in the biosynthesis of catecholamines (dopamine, noradrenaline and adrenaline)[1]. Catecholamines function as neurotransmitters in dopamine, noradrenaline and adrenaline neurons in the brain and retina, and in peripheral sympathetic noradrenaline neurons, and also as hormones (adrenaline and noradrenaline) in the adrenal medulla. Catecholamine neurotransmitters in the brain regulate a wide range of high-level brain functions, such as movement, emotion, learning, memory, biorhythm, reproduction and endocrine function, by acting across synapses through dopamine receptors and α- and β-adrenaline receptors of the neuronal network. In peripheral tissues, noradrenergic sympathetic neurons distributed in organs secrete noradrenaline as a neurotransmitter from their nerve endings. The adrenomedullary cells secrete adrenaline (and a small amount of noradrenaline) into the blood as hormones which regulate various functions indispensable to the maintenance of life, e.g. autonomic function, stress reactions, blood glucose level, blood pressure and blood circulation, by acting on cells with α- and β-adrenaline receptors.

TH plays important roles in physiology and pathology through the regulation of catecholamine biosynthesis. Catecholamines are known to be

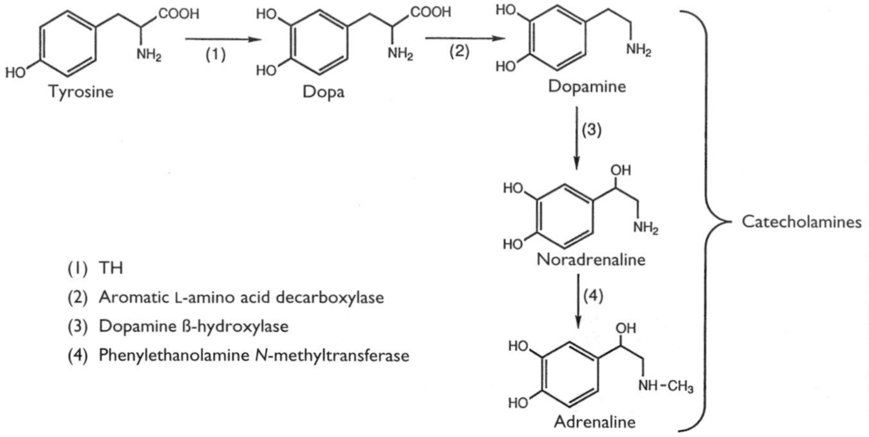

Figure 1. Pathway of biosynthesis of catecholamines (dopamine, noradrenaline and adrenaline) from tyrosine, and the catecholamine-synthesizing enzymes

involved in many diseases, including neuropsychiatric diseases (Parkinson's disease, affective disorders or manic depressive illness, schizophrenia, etc.); cardiovascular diseases (hypertension, cardiac diseases, etc.); and metabolic diseases (diabetes mellitus, etc.).

Catecholamines are synthesized from L-tyrosine by the pathway shown in Figure 1. Thus dopaminergic neurons contain the following synthesizing enzymes: (1) TH and (2) aromatic L-amino acid decarboxylase (EC 4.1.1.28; also known as dopa decarboxylase). Noradrenergic neurons or adrenomedullary cells have a third synthesizing enzyme: (3) dopamine β-hydroxylase (EC 1.14.17.1; also known as dopamine β-monooxygenase). Cells that synthesize adrenaline also have a fourth synthesizing enzyme in addition to these three: (4) phenylethanolamine N-methyltransferase (EC 2.1.1.28; also known as noradrenaline N-methyltransferase).

TH was discovered in 1964[1]. At that time, of the four enzymes involved in catecholamine biosynthesis, only the enzyme responsible for converting tyrosine to dopa was elusive. Tyrosinase was assumed to catalyse the reaction, but was not found in catecholamine-containing tissues, including the brain. Others had suggested that dopa formation was non-enzymic *in vivo*, since it could be observed easily under various conditions *in vitro*. TH activity was first detected with a newly developed, sensitive radio-isotopic assay which used L-[14C]-tyrosine as substrate. L-[14C]-dopa, enzymically formed, was isolated on an alumina column and assayed; however, when D-[14C]-tyrosine was used as a control, no radiolabelled dopa was formed. This evidence clearly demonstrated that an enzyme, such as TH, catalyses the conversion of L-tyrosine to L-dopa. TH was later found to be the rate-limiting enzyme in the biosynthesis of catecholamines[2].

TH requires a pteridine and ferrous ion as essential cofactors[1]. The natural tetrahydropteridine cofactor, tetrahydrobiopterin, was found to be most active[3]. The enzyme requires molecular oxygen as a substrate and is therefore a monooxygenase (also known as tyrosine 3-monooxygenase)[1].

The purification of TH was difficult, but was finally achieved in early 1980. By 1990, complex regulatory mechanisms had been found, including feedback regulation by catecholamines, and activation or deactivation due to phosphorylation by protein kinases or dephosphorylation by phosphatases. Since 1985, the structure of TH from various species, including humans, has been determined by cDNA cloning.

Properties of TH as a pteridine-dependent monooxygenase

TH is expressed in the catecholamine neurons that are present in discrete regions of the brain and retina, in the noradrenaline neurons of sympathetic ganglia and sympathetic nerves, and in adrenaline and noradrenaline cells of the adrenal medulla.

The reaction of TH is considered to be similar to that of phenylalanine 4-hydroxylase (EC 1.14.16.1)[4] (Figure 2). The TH substrates L-tyrosine and molecular oxygen and the tetrahydrobiopterin natural cofactor are converted to L-dopa and 4a-carbinolamine tetrahydrobiopterin. 4a-Carbinolamine tetrahydrobiopterin is converted to quinonoid dihydrobiopterin by pterin 4a-carbinolamine dehydratase (EC 4.2.1.96). Quinonoid dihydrobiopterin is

Figure 2. Reaction catalysed by TH in relation to the tetrahydrobiopterin cofactor

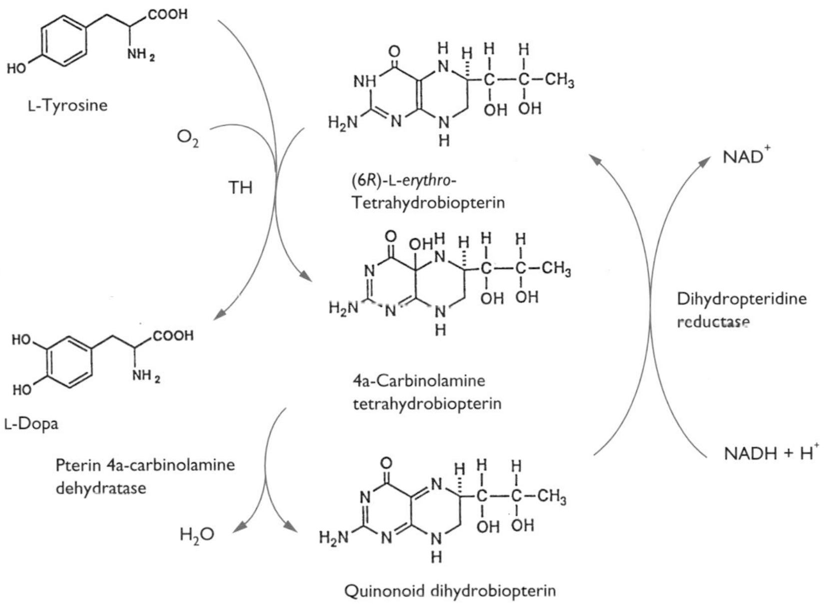

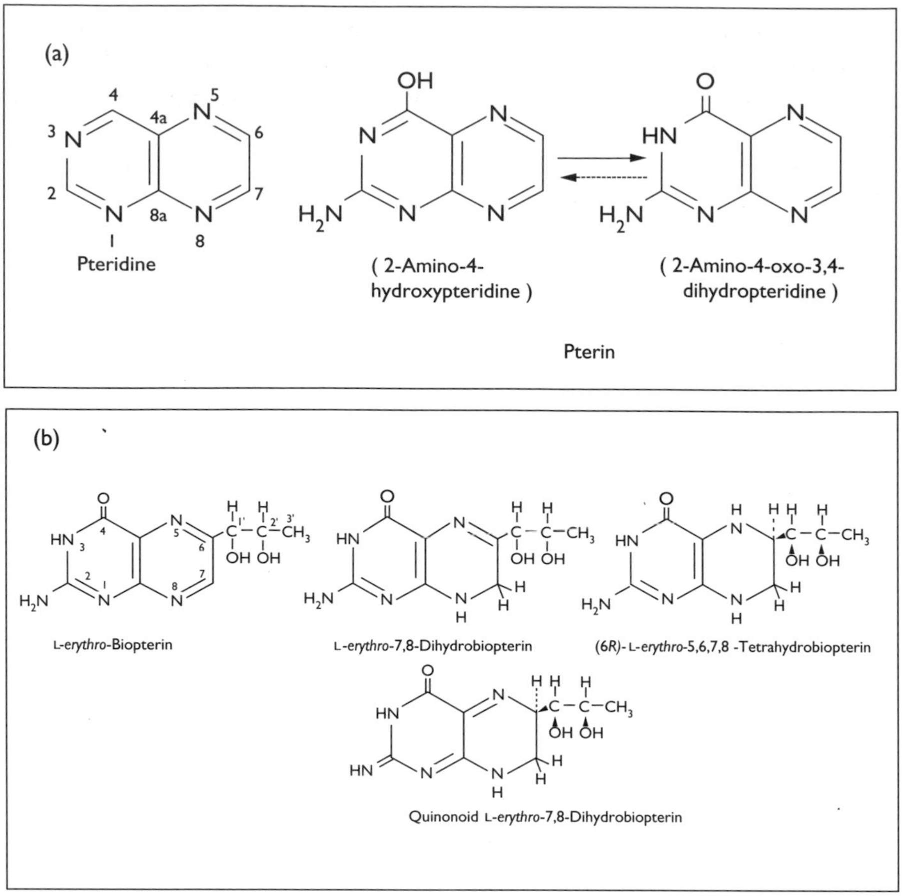

Figure 3. (a) Structure of pteridine and pterin; (b) structure of four forms of tetra-hydrobiopterin

reduced back to tetrahydrobiopterin by dihydropteridine reductase (EC 1.6.99.7) with NADH as the cofactor.

The term pterin, which was originally used to describe a factor in the pigments of butterfly wings, is now used for the natural pteridine compounds, most of which have the structure of 2-amino-4-hydroxypteridine. *In vivo*, pterin has the structure of the oxoform, 2-amino-4-oxo-3,4-dihydropteridine (Figure 3a). The natural form of the tetrahydropteridine cofactor, L-*erythro*-tetrahydrobiopterin, was first discovered as the cofactor of phenylalanine 4-hydroxylase (Figure 3b). The stereochemical structure is the (6R)-form in the reduced tetrahydro- form. Enzymically produced quinonoid dihydro-biopterin is also spontaneously and rapidly converted to 7,8-dihydrobiopterin, and then further oxidized to biopterin. The tissue concentration of the latter two forms is low compared with the former two reduced forms.

Tetrahydrobiopterin has many important functions as the cofactor of pterin-requiring monooxygenases and also of nitric oxide synthase.

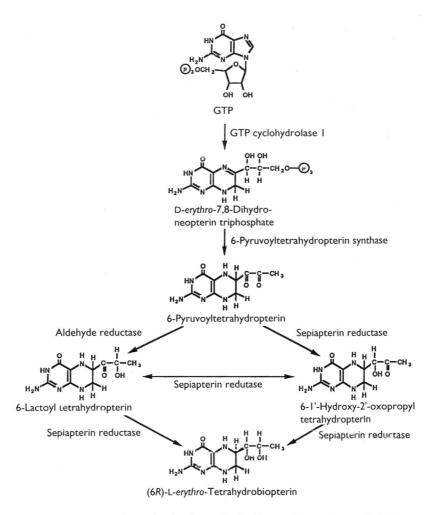

GTP

GTP cyclohydrolase I

D-*erythro*-7,8-Dihydro-
neopterin triphosphate

6-Pyruvoyltetrahydropterin synthase

6-Pyruvoyltetrahydropterin

Aldehyde reductase   Sepiapterin reductase

6-Lactoyl tetrahydropterin   Sepiapterin redutase   6-1'-Hydroxy-2'-oxopropyl
                                                                          tetrahydropterin

Sepiapterin reductase   Sepiapterin reductase

(6*R*)-L-*erythro*-Tetrahydrobiopterin

Figure 4. Pathway of biosynthesis of tetrahydrobiopterin, a cofactor for TH

Tetrahydrobiopterin is synthesized from GTP by the pathway shown in
Figure 4. Three enzymes are required: (i) GTP cyclohydrolase I (EC 3.5.4.16);
(ii) 6-pyruvoyltetrahydropterin synthase (EC 4.6.1.10); and (iii) sepiapterin
reductase (EC 1.1.1.153). The third step may be catalysed by sepiapterin
reductase alone or by aldehyde reductase (EC 1.1.1.21) and then by sepiapterin
reductase (Figure 5). The concentration of tetrahydrobiopterin synthesized
from GTP partly regulates the activity of TH.

TH also requires Fe^{2+} for activity. Human TH in crude tissue preparations
is highly activated by exogenously added Fe^{2+}.

TH has been purified from bovine adrenal medulla[5], rat adrenals[6], rat
pheochromocytoma[7], human adrenals and human brain[8]. In human adrenals
and brain, TH is composed of both active and less active forms. The less active
forms can be detected by enzyme immunoassay and Western blot analysis[7]. As

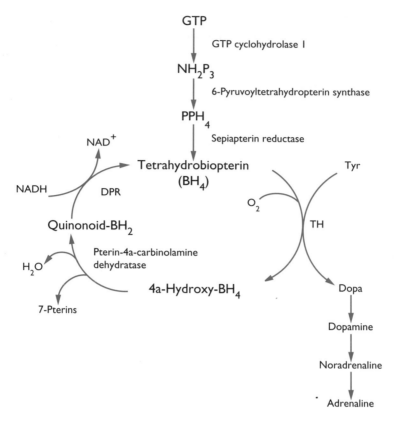

Figure 5. Relationship of tetrahydrobiopterin biosynthetic pathway to catecholamine biosynthesis via tyrosine hydroxylase
Abbreviations used: NH_2P_3, dihydroneopterin triphosphate; PPH_4, 6-pyruvoyltetrahydropterin; BH_4, tetrahydrobiopterin; BH_2, dihydrobiopterin; DPR, dihydropteridine reductase.

described below, human TH has four types of mRNA encoding four isoform proteins.

Rat, bovine or human TH is a tetrameric protein of about 240 kDa, each subunit having a mass of approx. 60 kDa. Each subunit has a C-terminal catalytic domain that binds the substrates tyrosine and molecular oxygen, and the pterin cofactor, and an N-terminal regulatory domain containing phosphorylated serine residues.

Isoforms of human TH

Since it was difficult to obtain sufficient amounts of TH protein to elucidate the complete amino acid sequence, the primary structure of TH from various species including humans has been deduced from the nucleotide sequence of TH cDNA. A full-length cDNA containing the entire sequence of rat TH was first cloned from rat pheochromocytoma[9]. The open-reading frame, including

```
                        .         .       1.| 3       .       .
Human    1   MPTPDATTPQAKGFRRAVSELDAKQAEAIMSPRFIGRRQSLIEDARKERE
Mouse    1   ----S-SS--P-----------Q-T-----VT--------------------
Rat      1   ----S-PS--P---------Q-------VT--------------------
Bovine   1   ----N-AS-------------------------V-------Q-----
Quail    1   ----NIS-SA------------S---------------------------

                        .         .         .         .         .
Human   51   AAVAAAAAAVPS EPGDPLEAVAFEEKEGKAVLNLLFSPRATKPSALSRA
Mouse   51   --A-------A-A---N-----V---RD-N--------L-G----S----
Rat     51   --A-------A-S---N-----V---RD-N--------L-G----S----
Bovine  51   K-E---SSS.......ESA--GSLV-RD-----T---AL-P---P--T--
Quail   51   --A--TD--.......EST-TIV----D-R-M---F-MLKGV-T-P----

               3 | 4  .         .         .         .         .
Human  100   VKVFETFEAKIHHLETRPAQRPRAGGPHLEYFVRLEVRRGDLAALLSGVR
Mouse  101   L-------------------L--S--------F--PS--------S--
Rat    101   --------------------L--S--------F--PS--------S--
Bovine  94   I--------HL---------PL---S-P--C---C--PGPVVP----AL-
Quail   94   L----------------LSRK--E-TAE------C--HSS--NTFI-SIK

                4 | 5.          .         .       . 5 | 6    .
Human  150   OVSEDVRSPAGPKVPWFPRKVSELDKCHHLVTKFDPDLDLDHPGFSDQVY
Mouse  151   R--D----ARED--------------------------------------
Rat    151   R--D----ARED-------------------------------------A
Bovine 144   R-A----AAGES--L---------------------------------A-
Quail  144   R-A----TTKED-FE-----IC---------------------Y-----

                       .    6 | 7 .          . 7 | 8    .         .
Human  200   RQRRKLIAEIAFQYRHGDPIPRVEYTKEEIATWKEVYTTLKGLYATHACG
Mouse  201   ---------------KQ-E---H---------------A-----------R
Rat    201   ---------------K--E---H----A----------V----------R
Bovine 194   ---------------K--E---H---------------V----------R
Quail  194   -----S------H-K--------------T-------S---S--P----K

                        .         .       8 | 9     .         .
Human  250   EHLEAFALLERFSGYREDNIPQLEDVSRFLKERTGFQLRPVAGLLSARDF
Mouse  251   ------Q----YC-----S--------H----------------------
Rat    251   ----G-Q----YC-----S-------------------------------
Bovine 244   ------E-----C-----R-------------------------------
Quail  244   -Y----N---K-C--N-N------E-----------------R--------

                        .         .     9 | 10          .     10.|
Human  300   LASLAFRVFQCTQYIRHASSPMHSPEPDCCHELLGHVPMLADRTFAQFSQ
Mouse  301   --------------------------------------------------
Rat    301   --------------------------------------------------
Bovine 294   ----------------------------------A-G-------------
Quail  294   -------------------------------------------K-------

              11        .       11 |.12         .         .      12.
Human  350   DIGLASLGASDEEIEKLSTLSWFTVEFGLCKQNGEVKAYGAGLLSSYGEL
Mouse  351   ---------------------VY---------------L--------------
Rat    351   ---------------------VY---------------L--------------
Bovine 344   ---------V-----------Y----------------N--------------
Quail  344   ----------T-------A--Y----------R---I----------------

              | 13       .         .         .         .   13 | 14
Human  400   LHCLSEEPEIRAFDPEAAAVQPYQDQTYQSVYFVSESFSDAKDKLRSYAS
Mouse  401   --S------V-----DT----------------P-------------------N---
Rat    401   --S--------V-----DT-------------P---------N---------N---
Bovine 394   --S--------------D-----------------P---------------------
Quail  394   I-S--D---V-D---D-------C---P--P---------------N---N--A

                        .         .         .
Human  450   RIQRPFSVKFDPYTLAIDVLDSPOAVRRSLEGVQDELDTLAHALSAIG 497
Mouse  451   -----------------------HTI-----------H--TQ-----S 498
Rat    451   -----------------------HTIQ----------H---------S 498
Bovine 444   -----------------------H-I--A-D-----MQA-----N--S 491
Quail  444   H-K------YE---HS-EL-----TICH---S-R---H--IN--NV-S 491
```

Figure 6. Comparison of the structures of human (type 1), mouse, rat, bovine and quail TH
Identical amino acids of mouse, rat, bovine and quail TH are expressed by hyphens. Vertical bars and the numbers above the human amino acid sequence represent break-point of exons and the exon numbers in the human TH gene, as shown in Figure 8.

the initiation codon, contains 1494 bp that encode 498 amino acids. Only one cDNA was cloned from rat, mouse, or bovine tissues. Figure 6 shows a comparison of amino acid sequences between human TH (type 1) and animal (mouse, rat, bovine and quail) TH. The sequence similarity of TH from various animals is high at the catalytic domain near the carboxyl region.

In contrast with a single TH cDNA in animals, human TH has four isoforms (hTH1–4) of mRNA encoding different proteins[10,11] (Figure 7). Nucleotide sequence analyses of full-length cDNA of types 1 and 2[10], type 3 and type 4[11] revealed that these four mRNAs differ only in the inclusion/exclusion of 12, 81 and 93 (12 plus 81) bp sequences, respectively, between nucleotides 90 and 91 of hTH1 mRNA. Since this insertion does not alter the reading frame of the protein-coding region, type-4 cDNA encodes the longest TH molecules. Southern blot analysis of human genomic DNA has

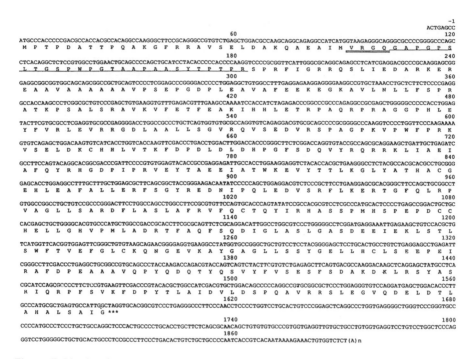

Figure 7. Nucleotide sequence and deduced amino acid sequence of hTH4 cDNA
The 81 bp sequence on the single line and the 12 bp sequence on the double line are deleted in hTH2 cDNA and hTH3 cDNA, respectively. The two sequences corresponding to the 93 bp are deleted in hTH1 cDNA, which is common among TH cDNA from various animals.

suggested that the human TH gene exists as a single gene per haploid DNA, indicating that these different human mRNAs are produced through alternative mRNA splicing from a single primary transcript[11].

Genomic clones encoding the human TH gene were isolated and characterized[12,13]. The human TH gene is composed of 14 exons, interrupted by 13 introns, spanning approximately 8.5 kb (Figure 8). The nucleotide sequence of the coding regions is the same as that of type-4 cDNA. The 12 bp insertion sequence is derived from the 3′-terminal portion of exon 1 (also called exon 1_2) and the 81-bp insertion sequence is encoded by exon 2 (also called exon 1_3). The N-terminal region is encoded by the 5′-portion of exon 1 (also called exon 1_1), and the remaining region from exon 3 to exon 14 (also called exon 2 to exon 13 for comparison with animal TH genes), is common to all four kinds of mRNA. Figure 8 summarizes the alternative splicing patterns which generate the four types of human TH mRNA. There are two modes of alternative splicing: (i) the alternate use of two donor sites in exon 1 (also called exon 1_1 and exon 1_2), whereby the selection of the two donor sites determines the insertion/deletion of the 12 bp sequence (also called exon 1_2); (ii) the other mode is the insertion/exclusion of an entire exon 2 (also called exon 1_3) that is specific for the human TH gene. Expression of type 1/2 or type 3/4 human TH

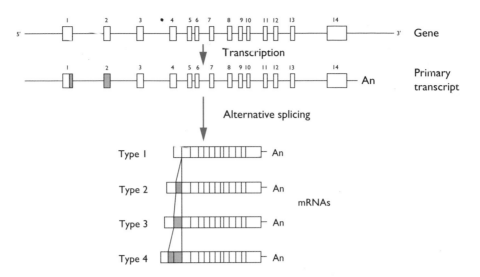

Figure 8. Structure of the human TH gene and schematic illustration of the alternative splicing pathway producing the four types of human TH mRNA from the primary transcript
The 3′-terminal, 12 bp sequence of exon I is also named exon I_1, and the exon 2 is also named exon I_3.

mRNA is determined by exclusion or inclusion of exon 2 (or exon 1_3) in the spliced products. The other 12 exons downstream from exon 3 (also called exon 2 for comparison with animal TH genes) are spliced and incorporated into mature mRNA.

hTH1 is similar to the enzyme from various animals. hTH1–4 have been expressed in COS cells, in *Xenopus* oocytes and in vertebrate cells. The expressed human TH types 1–4 show similar K_m values for tyrosine and the pteridine cofactor. However, the four types of human TH have different specific activities: hTH1 has the highest specific activity; the values for the other enzymes range from about 30% to 40% of that of hTH1.

hTH1–4 cDNAs have also been expressed in *Escherichia coli*, and large amounts of pure human TH have been obtained to characterize their properties[14,15].

mRNAs encoding the four isoforms of human TH have been detected in human neuroendocrine tissues and quantitatively determined in human brain (substantia nigra) using reverse transcription-polymerase chain reaction (RT-PCR). hTH1 and hTH2 are major species, and hTH3 and hTH4 are minor species. About 5% of the total human TH mRNA is represented by hTH3 and hTH4 in the normal human substantia nigra[16]. The approximate ratio of hTH1, hTH2, hTH3 and hTH4 mRNAs to the total amount of TH mRNA is 45:52:1.4:2.1[16].

TH isoform-specific, anti-oligopeptide antibodies were produced, and all four isoform proteins were detected in the human adrenal medulla and human brain[17]. The estimated ratio of isoforms in the human adrenal medulla was 40:40:10:7. Since hTH3 and hTH4 mRNA contents were higher in the adrenal medulla than in the brain, the ratio of hTH1–4 proteins is thought to be similar to the ratio of TH isoform mRNAs present.

TH isoforms are also found in monkeys[18,19]. Analyses of mRNA and/or genomic DNA of marmosets (New-world monkeys), crab-eating monkeys (*Macaca irus*), Japanese monkeys (*Macaca fuscata*, Old-world monkeys) and gorillas using PCR indicate that multiple types of TH corresponding to hTH1 and hTH2 are present, and that the isoforms corresponding to hTH3 and hTH4 are absent. Chimpanzee, orangutan and gibbon were also suggested to have types 1 and 2 from the genomic DNA sequences, but these higher apes, except gorilla, may also have the capacity to produce type-3 and type-4 mRNA. Direct analysis of mRNAs would be required to determine the existence of types 3 and 4 in these anthropoids. Immunohistochemical studies have revealed that both type-1 and type-2 TH proteins, but not types 3 or 4, are expressed in the brain of macaque monkeys[20]. These results indicate that New- and Old-world monkeys and gorillas produce TH types 1 and 2, and that mutations that had accumulated in the genomic DNA create a new exon (exon 2 or exon 1_3), resulting in the appearance of two new TH isoforms, types 3 and 4, in humans[19]. Phylogenetic trees of hominoids suggest that the gibbon split off from the common ancestor first, followed by the orangutan and gorilla. Finally, the chimpanzee and human separated about 5 million years ago. Distances between gorilla and human, and between chimpanzee and human, are very close. The increased heterogeneity of TH, from a single isoform in non-primate animals, to two isoforms in monkeys and four isoforms in humans, offers new insight into the sequence of events leading to the evolution into separate species of the high primates.

Generation of heterogeneity in the TH isoforms in monkeys and humans may alter the biosynthesis of catecholamines *in vivo*, and might affect the growth of neurites and the neural circuitry in the brain. Since TH regulates the biosynthesis of catecholamines that are essential for higher brain function, it is tempting to speculate that the genetic difference among humans, primates and non-primates is related to a specific brain function.

Regulation of TH

TH is regulated in a very complex way: in the short term, TH activity is mainly regulated by phosphorylation of serine residues in the regulatory domain at the N-terminus by various protein kinases; in the long term, such as under stress, TH is regulated at the transcriptional level resulting in the induction of TH (Figure 9).

TH purified from various species is a 240 kDa homotetramer composed of four subunits of approx. 60 kDa each. Each of the purified human TH types 1–4 expressed in *E. coli* also has a tetrameric structure. Limited proteolysis

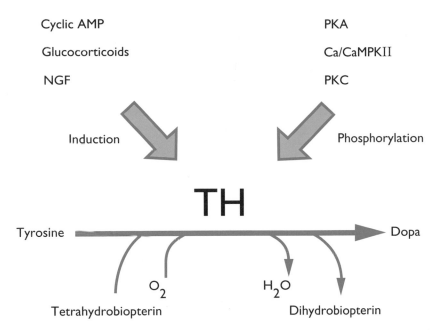

Figure 9. The mechanism of short-term regulation of TH by phosphorylation (activation) and long-term regulation by induction (gene expression)
Protein kinases (PKA, Ca/CaMPKII and PKC) activate TH via phosphorylation in the short term and also induce TH via gene activation in the long term.

reveals the inhibitory regulatory domain at the N-terminus and the catalytic domain at the C-terminus. Deletion mutagenesis studies place the C-terminal catalytic domain of rat TH between residues 158 and 184, and the carboxyl end at or prior to position 455[21]. Mature TH purified from adrenals or brain, or recombinant TH expressed in *E. coli*, exists as a homotetramer. A region containing a putative C-terminal leucine zipper may be required for TH tetramer formation[22].

Figure 10 shows a schematic presentation of the short-term regulation of dopamine biosynthesis via regulation of TH activity in the dopaminergic nerve terminals in the basal ganglia of the brain. The concentration of the cofactor, tetrahydrobiopterin, is a regulatory factor. TH is not saturated with tetrahydrobiopterin *in vivo*, and the cofactor level which is mainly regulated by GTP cyclohydrolase I activity may also regulate TH activity.

Catecholamines, the end product of the TH reaction, inhibit the enzyme activity competitively with tetrahydrobiopterin[1], and inactivate the enzyme reversibly to convert the active/labile form to an inactive/stable form[23]. These two feedback inhibition mechanisms by catecholamines are important in short-term regulation. Bovine adrenal TH is isolated in the inhibited state with catecholamines as the blue-green coloured catecholamine–Fe^{2+} complex. Phosphorylation of Ser-40 at pH 7.0 causes the release of catecholamine to activate the enzyme[24].

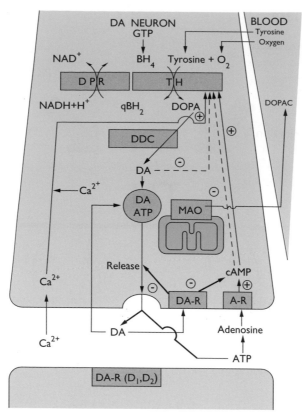

Figure 10. Schematic presentation of the short-term regulation of dopamine bio-synthesis via TH activity in the dopaminergic nerve terminals in the basal ganglia of the brain

Abbreviations used: A-R, adenosine receptor; BH_4, tetrahydrobiopterin; cAMP, cyclic AMP; DA, dopamine; DA-R, dopamine receptor; DDC, dopa decarboxylase (AADC, aromatic L-amino acid decarboxylase); DOPA, L-dopa; DOPAC, 3,4-dihydroxyphenylacetic acid; DPR, dihydropteridine reductase; MAO, monoamine oxidase; qBH_2, quinonoid dihydrobiopterin.

Another probable regulation of TH activity is activation by association with chromaffin granules in the adrenal medulla or with synaptic vesicles. TH from the cytosol can bind reversibly to the granule membrane in a process that results in activation. This attachment of TH on the surface of chromaffin granules has been confirmed by immuno-electronmicroscopy[25].

The most important short-term mechanism for regulation of TH is activation by phosphorylation via protein kinases and deactivation by dephosphorylation via protein phosphatases. As shown in the schematic diagram of Figure 11, the main phosphorylation sites of TH in vitro are Ser-19, Ser-31 and Ser-40[26]. Ser-19 is phosphorylated mainly by Ca^{2+}/calmodulin-dependent protein kinase II (Ca/CaMPKII; EC 2.7.1.123), while Scr-40 is phosphorylated mainly by protein kinase A (PKA). Ca/CaMPKII may phosphorylate and activate TH of PC12h cells when they are depolarized by high K^+ because

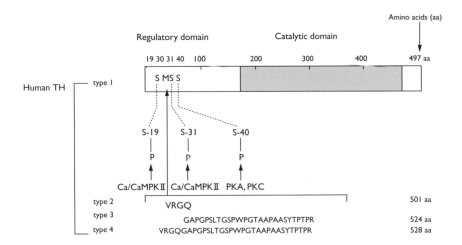

Figure 11. Schematic diagram of the main phosphorylation sites (Ser-19, Ser-31 and Ser-40) of hTH1
The open area shows the regulatory N-terminal domain. The shaded area shows the catalytic C-terminal domain. S-19, S-31 and S-40 are the main phosphorylation sites activating the enzyme. The insertion sequences of 4, 27 and 31 amino acids between M-30 and S-31 correspond to hTH2, hTH3 and hTH4.

a selective inhibitor of Ca/CaMPKII, KN-62 {1-[N,O-bis(5-isoquinolin-sulfonyl)-N-methyl-L-tyrosyl]-4-phenylpiperazine}, inhibits this TH phosphorylation and reduces dopamine synthesis. These results agree with the report that Ca/CaMPKII mediates phosphorylation of TH by hormonal and electrical stimuli, which leads to elevation of Ca^{2+} levels in PC12 cells[27]. Ser-40 is phosphorylated by PKA, protein kinase C (PKC) and Ca/CaMPKII; however, PKA phosphorylates Ser-40 of all four subunits of the enzyme molecule, causing a marked activation, whereas PKC and Ca/CaMPKII phosphorylate only two of the four subunits without affecting the enzyme activity[28]. Ser-31 is also phosphorylated by the extracellular signal-regulated kinases 1 and 2 (ERK1 and ERK2), two microtubule-associated protein kinases[29].

The first messengers likely to regulate TH phosphorylation include: dopamine (via the presynaptic dopamine autoreceptor); adenosine (via the presynaptic A2 receptor); glutamate [via the N-methyl-D-aspartate (NMDA) receptor]; vasoactive intestinal polypeptide (VIP) (via PKA); angiotensin (via PKC); secretin–glucagon (via PKA); prolactin (via PKC); and nerve growth factor (NGF) (via Ca/CaMPKII). Dephosphorylation of TH by protein phosphatases (type 2A) decreases the activity. However, the enzyme expressed in *E. coli* has high activity without phosphorylation, indicating that the unphosphorylated enzyme has activity. Another finding, suggesting the regulatory inter-relationship between the tetrahydrobiopterin synthetic pathway and catecholamine biosynthesis, is that tetrahydrobiopterin activates TH phosphatase.

Thus increased concentrations of tetrahydrobiopterin activate TH, but may also decrease its activity due to dephosphorylation[30].

The mechanism of activation of TH by phosphorylation at Ser-40 is increased affinity for the tetrahydrobiopterin cofactor and removal of inhibition by the end-product catecholamine[31]. The insertion sequence between Met-30 and Ser-31 of hTH1 promotes additional phosphorylation of hTH2 by Ca/CaMPKII. Unlike hTH1, phosphorylation of hTH2 by Ca/CaMPKII results in an increase of the K_i value for dopamine, giving a greater potential for activation than hTH1. The hTH1–4 isoforms are phosphorylated at Ser-40 and Ser-19 by mitogen-activated protein-kinase (MAP kinase)-activated kinase-1 and -2 (MAPKAP kinase-1 and -2), and at Ser-31 by MAP kinase. It is suggested that phosphorylation by MAPKAP kinase-1 and -2 may be of particular importance for the regulation of hTH2, which is phosphorylated by MAP kinase very poorly, and that phosphorylation by MAP kinase may be of special significance for the regulation of hTH3 and hTH4[32].

hTH1, hTH2 and hTH4 are inhibited by catecholamines in competition with tetrahydrobiopterin. Catecholamines bind to hTH1 and hTH2 with a stoicheiometry of about 1 mol per mol of enzyme subunit interacting with the catalytic iron at the active site. Tetrahydrobiopterin causes a dissociation of dopamine from hTH1. Phosphorylation at Ser-40 by PKA decreases the affinity of dopamine binding by a factor of 10. These results suggest that human TH isoforms are regulated in a similar fashion to TH from other species.

Gene expression of TH

TH is regulated in the long term, such as under chronic stress, by enzyme induction at the transcriptional level. As shown in Figure 12, several putative regulatory elements exist in the 5′-upstream region of the genes of human and rat TH within 0.2 kb of the 5′-flanking DNA sequence: AP2, AP1, POU/Octa, Hepta, Sp1 and cyclic AMP response element (CRE).

Protein kinases (PKA, Ca/CaMPKII, PKC etc.) activate TH by phosphorylation in the short term, and also induce TH protein in the long term. Thus protein kinases have dual regulatory roles.

The expression of TH in cultured cells and tissues containing catecholamines is regulated by various first messengers (e.g. dopamine, dopamine

Figure 12. A schematic map of the 5′ upstream region of the human TH gene

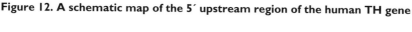
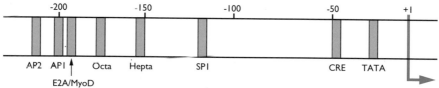

agonists and antagonists, dexamethasone, VIP and secretin, angiotensin II, bradykinin, neurotensin and NGF), and by PKA and PKC signal transduction pathways.

Functional CRE activity has been found in a variety of cell lines. The CRE appears to play an important dual role: as a basal promoter element and an inducible enhancer for TH transcription. CRE and CRE-binding protein (CREB) may play a fundamental role in the transcriptional activation of the TH gene in catecholaminergic cells[33,34].

The AP1 site may also functionally regulate TH gene activity, but may contribute to transcription to a smaller degree than CRE. Binding of the products of immediate early genes, c-Fos- and c-Jun-related proteins induced by NGF or angiotensin II, to the putative AP1-like sites increases TH transcription. Thus NGF treatment of responsive cells increases TH action by two different mechanisms: the first is a short-term elevation of TH activity due to an increase in TH phosphorylation; the second is a long-term elevation of TH due to an increase in the synthesis of the enzyme. Both PKA and PKC appear both to activate TH by phosphorylation and to induce its synthesis by an increase in TH transcription. It has also been proposed that the tissue-specific regulation of TH requires a synergistic interaction between the AP1 motif and the overlapping E-box.

Cold-induced increases in adrenomedullary TH gene expression are mediated through the interaction of the AP1 binding site and the c-Fos/c-Jun. Membrane depolarization induces an increase in intracellular Ca^{2+} which in turn induces TH. The depolarization response element in the TH gene in PC12 cells is thought to be CRE. Thus CRE appears to be functioning as a calcium regulatory element in this system.

Analysis of TH gene expression in transgenic mice

Since the expression pattern of TH is spatially and temporally specific, transgenic mice are useful for defining the regulation elements for TH gene expression. The transgenic (HTH) mice carrying an 11 kb fragment (containing a 2.5 kb 5′-flanking region, the entire exon–intron sequence and a 0.5 kb 3′-flanking region) exhibited high-level and tissue-specific expression of human TH in the brain and adrenal glands[35]. The 5.0 kb 5′-flanking region of the human TH gene could drive chloramphenicol acetyltransferase (CAT) reporter gene expression in catecholaminergic neurons and adrenal medullary cells of non-transgenic mice; however, CAT expression was also observed in some non-catecholaminergic neurons, including those in several sites where transient TH expression has been reported. The 2.5 kb and 0.2 kb 5′-flanking fragments of the TH gene could not express CAT in catecholaminergic neurons[36]. The 5.0 kb of the human TH 5′-flanking region, the exon–intron structure and/or 3′-flanking region of the TH gene may function in catecholaminergic neuron-specific expression. The results in HTH transgenic mice show that the fundamental cellular machinery necessary for the alternative splicing of human TH mRNA is present and functioning in the mouse catecholaminergic cells and produces multiple forms of the enzyme from human TH mRNA sequence. Although human TH mRNA and active protein are overexpressed in the HTH transgenic mice, the catecholamine levels and phenotypes are not significantly different from those of non-transgenic mice, suggesting that there are other, unknown, regulatory mechanisms for the catecholamine levels in the transgenic mice. In transgenic mice, introducing either 4.8 kb or 9 kb of the 5′ flanking region of the rat TH gene is sufficient for the high level of tissue-specific expression[37,38]. Thus it may also be possible that other catecholaminergic neuron-specific elements reside between 5 kb and 9 kb of the human TH gene, as well as in the intron–exon structure and/or the 3′-terminal region.

TH in disease

Since catecholamines are closely related to the pathogenesis of neuropsychiatric or cardiovascular disorders, TH has been suggested to play an important role in a number of diseases.

In Parkinson's disease, TH activity, protein levels and mRNA levels are decreased in the nigrostriatal dopaminergic neurons[39]. A quantitative RT-PCR method for the four types of human TH mRNA revealed that the four isoforms exist in the human substantia nigra at an approximate ratio of 45:50:2:3. In Parkinsonian substantia nigra, each form of TH mRNA was decreased to about 25% of the normal level (Figure 13). In contrast, neither the absolute amount nor the ratio of hTH1–4 protein changed in schizophrenia. If hTH1–4 mRNAs exist in the same neuron, the total amount but not the ratio of hTH1–4 may change. On the other hand, neurons containing only one type

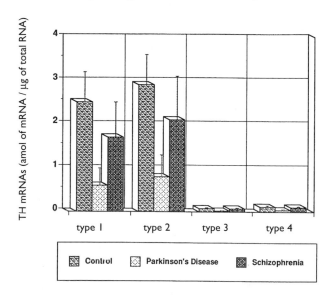

Figure 13.
Quantification of mRNAs of human TH isoforms in the substantia nigra[16]
Total amounts of TH mRNA (amol TH mRNA/μg of total RNA) in control, Parkinson's disease and schizophrenia were 5.4 ± 1.4, 1.5 ± 0.9 and 4.0 ± 1.8, respectively. The total amount, type 1 and 2 mRNAs were significantly reduced compared with corresponding control values ($P<0.05$).

of human TH protein were demonstrated by immunohistochemistry[17]. It is interesting that the surviving dopaminergic neurons in Parkinson's disease have decreased levels of TH mRNA and protein, suggesting the incapability of the remaining neurons[40].

The nigrostriatal dopamine neurons appear to be the most susceptible to dopamine deficiency. The first symptom (phenotype) of dopamine deficiency may be dystonia, i.e. disordered tonicity of muscle. Hereditary progressive dystonia with marked diurnal fluctuation (HPD) (also called dopa-responsive dystonia, DRD) is a dystonia with autosomal dominant inheritance with dopamine deficiency in the nigro-striatum of the brain, originally described by Segawa and sometimes known as Segawa's syndrome. Small doses of L-dopa can cure the patients. HPD/DRD is caused by mutation of GTP cyclohydrolase I, the first enzyme in tetrahydrobiopterin biosynthesis. The resultant decrease in tetrahydrobiopterin, to below 20% of the normal level, causes the decreased TH activity and dopamine deficiency[41]. A recessive inherited form of HPD/DRD in other families is caused by point mutation of TH (Gln-381 → Lys)[42]. It should be noted that another autosomal recessive condition of GTP cyclohydrolase I deficiency is also caused by a point mutation, resulting in no enzyme activity and severe neurological symptoms[43].

Catecholamine neurotransmitters are assumed to be closely related to mental diseases, especially to bipolar affective disorders (manic depressive illness) or schizophrenia. The TH gene is a part of a gene cluster of TH–insulin gene–insulin-like growth factor 2 gene in human chromosome 11p15.5. The TH gene is 5′ to the insulin gene and is separated by only 2.7 kb of flanking DNA. Since the first report in 1981 on linkage study, suggesting association between TH and the bipolar affective disorders, extensive studies have been carried out with conflicting results. A positive association between the bipolar illness and a locus containing the gene for TH has been reported[44]. Increased TH activity is assumed to explain probable overactivity of ventro-tegmental dopaminergic neurons in schizophrenia, but the amounts of hTH1–4 mRNAs in the substantia nigra do not change from those from normal controls[16]. It is obviously necessary to carry out molecular genetic studies with increased numbers of patients and families with affective disorders or schizophrenia.

An interesting therapeutic approach for Parkinson's disease is brain transplantation of non-neuronal cells transfected with human TH gene. This work is still at the level of animal experimentation. One important factor is the availability of the cofactor of TH, tetrahydrobiopterin[45], since tetrahydrobiopterin is essential for TH activity. Use of an adenovirus vector for gene transfer of TH into the substantia nigra by stereotaxic inoculation may be a promising approach as a gene therapy for Parkinson's disease.

Summary

- *TH is a tetrahydrobiopterin-requiring, iron-containing monooxygenase. It catalyses the conversion of L-tyrosine to L-dopa, which is the first, rate-limiting step in the biosynthesis of catecholamines (dopamine, noradrenaline and adrenaline), the central and sympathetic neurotransmitters and adrenomedullary hormones. The cofactor of TH is tetrahydrobiopterin, which is synthesized from GTP in three steps.*

- *The TH gene consists of 14 exons only in humans and 13 exons in animals. Human TH exists in four isoforms (hTH1–4) that are produced by alternative mRNA splicing from a single gene. A single mRNA and protein corresponding to hTH1 exists in non-primates. Monkey TH exists in two isoforms, corresponding to hTH1 and hTH2.*

- *TH activity is regulated in the short term by feedback inhibition of catecholamines in competition with tetrahydrobiopterin, and by activation and deactivation due to phosphorylation and dephosphorylation, mainly at Ser-19 and Ser-40 of hTH1. The multiple TH isoforms in humans and monkeys have additional phosphorylation, resulting in more subtle regulation.*

- *In long-term regulation under stress conditions, TH protein is induced. CRE and AP1 in the 5′ flanking region of the TH gene may be the main functional elements for TH gene expression.*

- *TH may be closely related to the pathogenesis of neurological diseases, such as dystonia and Parkinson's disease, psychiatric diseases, such as affective disorders and schizophrenia, as well as cardiovascular diseases.*

- *The TH gene may prove useful in gene therapy to compensate for decreased levels of catecholamines in neurological diseases, for example, for supplementation of dopamine in Parkinson's disease.*

I apologize to the many contributors to the field to whom I have not been able to refer, owing to the limitation of reference numbers. I thank the Ministry of Education, Science and Culture of Japan, and Fujita Health University for their support.

Further reading

Nagatsu, T. (1991) Genes for human catecholamine-synthesizing enzymes. *Neurosci. Res.* **12**, 315–345

Naoi, M. & Parvez, S.H., eds. (1993) *Tyrosine Hydroxylase*, VSP, Utrecht

Kaufman, S., ed. (1985) Metabolism of aromatic amino acids and amines. *Methods Enzymol.* **142**, Academic Press, New York

Zigmond, R.E., Schwartzschild, M.A. & Rittenheuse, A.R. (1989) Acute regulation of tyrosine hydroxylase by nerve activity and by neurotransmitters via phosphorylation. *Annu. Rev. Neurosci.* **12**, 415–461

References

1. Nagatsu, T., Levitt, M. & Udenfriend, S. (1964) Tyrosine hydroxylase: the initial step in norepinephrine biosynthesis. *J. Biol. Chem*, **239**, 2910–2917

2. Levitt, M., Spector, S., Sjoerdsma, A. & Udenfriend, S. (1965) Elucidation of the rate-limiting step in norepinephrine biosynthesis in the perfused guinea-pig heart. *J. Pharmacol. Exp. Therap.* **148**, 1–8

3. Brenneman, A.R. & Kaufman, S. (1964) The role of tetrahydropteridines in the enzymatic conversion of tyrosine to 3,4-dihydroxyphenylalanine. *Biochem. Biophys. Res. Commun.* **17**, 177–183

4. Lazarus, R., Benkovic, S. & Kaufman, S. (1983) Phenylalanine hydroxylase stimulator protein is a 4a-carbinolamine dehydratase. *J. Biol. Chem.* **238**, 10960–10962

5. Nagatsu, T. & Oka, K. (1987) Tyrosine 3-monooxygenase from bovine adrenal medulla. *Methods Enzymol.* **142**, 56–62

6. Fujisawa, H. & Okuno, S. (1987) Tyrosine 3-monooxgenase from rat adrenals. *Methods Enzymol.* **142**, 63–71

7. Tank, A.W. & Weiner, N. (1987) Tyrosine 3-monooxygenase from rat pheochromocytoma. *Methods Enzymol.* **142**, 71–82

8. Mogi, M., Kojima, K. & Nagatsu, T. (1984) Detection of inactive or less active forms of tyrosine hydroxylase in human brain and adrenal by a sandwich enzyme immunoassay. *Anal. Biochem.* **138**, 125–132

9. Grima, B., Lamouroux, A., Blanot, F., Biguet, N.F. & Mallet, J. (1985) Complete coding sequence of rat tyrosine hydroxylase mRNA. *Proc. Natl. Acad. Sci. U.S.A.* **82**, 617–621

10. Grima, B., Lamouroux, A., Boni, C., Julien, J.-F., Javoy-Agid, F. & Mallet, J. (1987) A single human gene encoding multiple tyrosine hydroxylases with different predicted functional characteristics. *Nature (London)* **326**, 707–711

11. Kaneda, N., Kobayashi, K., Ichinose, H., *et al.* (1987) Isolation of a novel cDNA clone for human tyrosine hydroxylase: alternative RNA splicing produces four kinds of mRNA from a single gene. *Biochem. Biophys. Res. Commun.* **146**, 971–975

12. O'Malley, K.L., Anhalt, M.J., Martin, B.M., Kalsoe, J.R., Winfield, S.L. & Ginns, E.I. (1987) Isolation and characterization of the human tyrosine hydroxylase gene: identification of 5′ alternative splice sites responsible for multiple mRNAs. *Biochemistry* **26**, 6910–6914

13. Kobayashi, K., Kaneda, N., Ichinose, H., *et al.* (1988) Structures of the human tyrosine hydroxylase gene: alternative splicing from a single gene accounts for generation of four mRNA subtypes. *J. Biochem.* **103**, 907–912

14. Le Bourdellès, B., Horellou, P., Le Caer, J.-P., *et al.* (1991) Phosphorylation of human recombinant tyrosine hydroxylase isoforms 1 and 2: an additional phosphorylated residue in isoform 2, generated through alternative splicing. *J. Biol. Chem.* **266**, 17124–17130

15. Nasrin, S., Ichinose, H., Hidaka, H. & Nagatsu, T. (1994) Recombinant human tyrosine hydroxylase types 1–4 show regulatory kinetic properties for the natural (6R)-tetrahydrobiopterin cofactor. *J. Biochem.* **116**, 393–398

16. Ichinose, H., Ohye, T., Fujita, K., *et al.* (1994) Quantification of mRNA of tyrosine hydroxylase and aromatic L-amino acid decarboxylase in the substantia nigra in Parkinson's disease and schizophrenia. *J. Neural Transm. [P-D Sect]* **8**, 149–158

17. Lewis, D.A., Melchitzky, D.S. & Haycock, J.W. (1993) Four isoforms of tyrosine hydroxylase are expressed in human brain. *Neuroscience* **54**, 477–492

18. Ichikawa, S., Icnihose, H. & Nagatsu, T. (1990) Multiple mRNAs of monkey tyrosine hydroxylase. *Biochem. Biophys. Res. Commun.* **173**, 1331–1336

19. Ichinose, H., Ohye, T., Fujita, K., Yoshida, M., Ueda, S. & Nagatsu, T. (1993) Increased heterogeneity of tyrosine hydroxylase in humans. *Biochem. Biophys. Res. Commun.* **195**, 158–165

20. Lewis, D.A. Melchitzky, D.S. & Haycock, J.W. (1994) Expression and distribution of two isoforms of tyrosine hydroxylase in macaque monkey brain. *Brain Res.* **656**, 1–13

21. Ribeiro, P., Wang, Y., Citron, B.A. & Kaufman, S. (1993) Deletion mutagenesis of rat PC12 tyrosine hydroxylase regulatory and catalytic domains. *J. Mol. Neurosci.* **4**, 125–139

22. Vrana, K.E., Walker, S.J., Rucker, P. & Liu, X. (1994) A carboxyterminal leucine zipper is required for tyrosine hydroxylase tetramer formation. *J. Neurochem.* **63**, 2014–2020

23. Okuno, S. & Fujisawa, H. (1991) Conversion of tyrosine hydroxylase to stable and inactive form by the end product. *J. Neurochem* **57**, 53–60

24. Andersson, K.K., Cox, D.D., Que, L., Jr., Flatmark, T. & Haavik, J. (1988) Resonance raman studies on the blue-green-coloured bovine adrenal tyrosine 3-monooxygenase (tyrosine hydroxylase). Evidence that the feedback inhibitors adrenaline and noradrenaline are coordinated to iron. *J. Biol. Chem.* **263**, 1821–18626

25. Nagatsu, I., Karasawa, N., Kondo, Y. & Inagaki, S. (1979) Immunocytochemical localization of tyrosine hydroxylase, dopamine-β-hydroxylase and phenylethanolamine-N-methyltransferase in the peroxidase-antiperoxidase method. *Histochemistry* **64**, 131–144

26. Campbell, D.G., Hardie,D.G. & Vulliet, P.R. (1986) Identification of four phosphorylation sites in the N-terminal region of tyrosine hydroxylase. *J. Biol. Chem.* **261**, 10489–10492

27. Ishii, A., Kiuchi, K.,Kobayashi, R., Sumi, M., Hidaka, H. & Nagatsu, T. (1991) A selective Ca^{2+}/calmodulin-dependent protein kinase II inhibitor, KN-62, inhibits the enhanced phosphorylation and the activity of tyrosine hydroxylase by 56 mM K^+ in rat pheochromocytoma PC12h cells. *Biochem. Biophys. Res. Commun* **176**, 1051–1056

28. Funakoshi, H., Okuno, S. & Fujisawa, H. (1991) Different effects on activity caused by phosphorylation of tyrosine hydroxylase at serine 40 by three multifunctional protein kinases. *J. Biol. Chem.* **266**, 15614–15620

29. Haycock, J.W., Ahn, N.G., Cobb, M.H. & Krebs, E.G. (1992) ERK1 and ERK2, two microtubule-associated protein 2 kinases, mediate the phosphorylation of tyrosine hydroxylase at serine-31 *in situ*. *Proc. Natl. Acad. Sci. U.S.A.* **89**, 2365–2369

30. Nelson, T.J. & Kaufman, S. (1987) Activation of rat caudate tyrosine hydroxylase phosphatase by tetrahydrobiopterin. *J. Biol. Chem.* **262**, 16470–16475

31. Bailey, S.W., Dillard, S.B., Thomas, K.B. & Ayling, J.E. (1987) Changes in the cofactor binding domain of bovine striatal tyrosine hydroxylase at physiological pH upon cAMP-dependent phosphorylation mapped with tetrahydrobiopterin analogues. *Biochemistry* **28**, 494–504

32. Sutherland, D., Alterio, J., Campbell, D.G., *et al.* (1993) Phosphorylation and activation of human tyrosine hydroxylase *in vitro* by mitogen-activated protein (MAP) kinase and MAP-kinase activated kinases 1 and 2. *Eur. J. Biochem.* **217**, 715–722

33. Cambi, F., Fung, B. & Chikaraishi, D. (1989) 5′ Flanking DNA sequences direct cell-specific expression of rat tyrosine hydroxylase. *J. Neurochem.* **53**, 1656–1659

34. Kim, K.-T., Park, D.H. & Joh, T.H. (1993) Parallel up-regulation of catecholamine biosynthetic enzymes by dexamethasone in PC12 cells. *J. Neurochem.* **60**, 946–951

35. Kaneda, N., Sasaoka, T., Kobayashi, K., *et al.* (1991) Tissue-specific and high-level expression of the human tyrosine hydroxylase gene in transgenic mice. *Neuron* **6**, 583–594

36. Sasaoka T., Kobayashi, K., Nagatsu, I., *et al.* (1992) Analysis of the human tyrosine hydroxylase promoter-chloramphenicol acetyltransferase chimeric gene expression in transgenic mice. *Mol. Brain. Res.* **16**, 274–286

37. Banerjee, S.A., Roffler-Tarlov, S., Szabo, M., Frohman, L. & Chikaraishi, D.M. (1994) DNA regulatory sequences of the rat tyrosine hydroxylase gene direct correct catecholaminergic cell-type specificity of a human growth hormone receptor in the CNS of transgenic mice causing a dwarf phenotype. *Mol. Brain Res.* **24**, 89–106.

38. Min, N., Joh, T.H., Kim, K.S., Peng, C. & Son, J.H. (1994) 5′ Upstream DNA sequence of the rat tyrosine hydroxylase gene directs high-level and tissue-specific expression to catecholaminergic neurons in the central nervous system of transgenic mice. *Mol. Brain Res.* **27**, 281–289

39. Nagatsu, T. (1990) Change of tyrosine hydroxylase in the parkinsonian brain and in the brain of MPTP-treated mice as revealed by homospecific activity. *Neurochem. Res.* **15**, 425–429

40. Kastner, A., Hirsch, E.C., Agid, Y. & Javoy-Agid, F. (1993) Tyrosine hydroxylase and messenger RNA in the dopaminergic nigral neurons of patients with Parkinson's disease. *Brain Res.* **606**, 341–345

41. Ichinose, H., Ohye, T., Takahashi, T., *et al.* (1994) Hereditary progressive dystonia with marked diurnal fluctuation caused by mutations in the GTP cyclohydrolase I gene. *Nature Genet.* **8**, 236–242

42. Lüdecke, B., Dworniezak, B. & Bartholomé, K. (1995) A point mutation in the tyrosine hydroxylase gene associated with Segawa's Syndrome. *Human Genet.* **95**, 123–125

43. Blau, N., Ichinose, H., Nagatsu, T., Heizmann, C.W., Zacchello, F. & Burline, A.B. (1995) A missense mutation in a patient with GTP cyclohydrolase I deficiency missed in the newborn screening program. *J. Pediatr.*, **126**, 401–405

44. Leboyer, M., Malafosse, A., Boularands, S., *et al.* (1990) Tyrosine hydroxylase polymorphisms associated with manic-depressive illness. *Lancet* **335**, 1219

45. Uchida, K., Tsuzaki, N., Nagatsu, T. & Kohsaka, S. (1992) Tetrahydrobiopterin-dependent functional recovery in 6-hydroxydopamine-treated rats by intracerebral grafting of fibroblasts transfected with tyrosine hydroxylase cDNA. *Devel. Neurosci.* **14**, 173–180

3

Antizyme-dependent degradation of ornithine decarboxylase

Shin-ichi Hayashi

Department of Nutrition, The Jikei University School of Medicine, Minato-ku, Tokyo 105, Japan

Introduction

Intracellular protein degradation is as important as protein synthesis for the modulation of the cellular levels of regulatory proteins, such as key enzymes of metabolism and cell-cycle regulators[1]. The process requires energy and is highly selective; the half-life of these regulatory proteins is very short (minutes) compared with the half-life of cellular proteins in general (days). Rapid turnover rates allow rapid changes in the cellular levels of regulatory proteins by changes in their synthesis rates. In some cases, the degradation rates themselves are regulated. These generalized comments are all applicable to the degradation of ornithine decarboxylase (ODC; EC 4.1.1.17), which has recently been clarified extensively at the molecular level.

ODC is a key enzyme in the synthetic pathway of polyamines, namely putrescine (a diamine), spermidine and spermine, as shown in Scheme 1[2]. ODC has the shortest half-life of all known mammalian enzymes and is dramatically induced by various kinds of growth stimulus[2,3]. On the other hand, ODC is repressed by polyamines for the protection of cells from the adverse effects of excess polyamines[4]. The degradation of ODC has been shown to be markedly accelerated by the binding of antizyme, an ODC inhibitory protein induced by polyamines[5,6], and to be catalysed by the 26S proteasome, a multicatalytic protease complex, in an ATP- and antizyme-dependent but ubiquitin-independent manner[7-9]. This is a novel type of negative-feedback

37

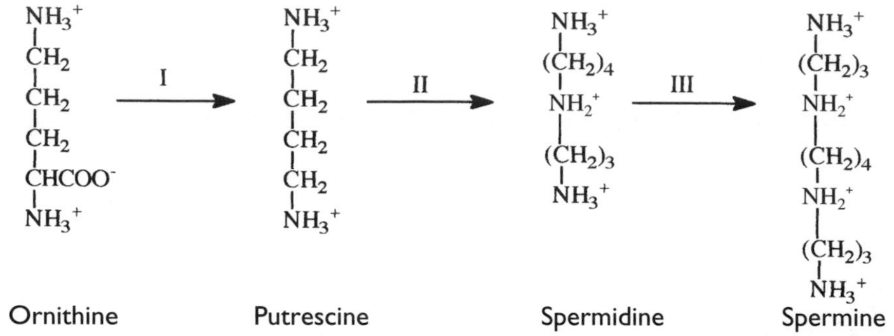

$$
\begin{array}{ccccccc}
NH_3^+ & & NH_3^+ & & NH_3^+ & & NH_3^+ \\
| & & | & & | & & | \\
CH_2 & & CH_2 & & (CH_2)_4 & & (CH_2)_3 \\
| & I & | & II & | & III & | \\
CH_2 & \longrightarrow & CH_2 & \longrightarrow & NH_2^+ & \longrightarrow & NH_2^+ \\
| & & | & & | & & | \\
CH_2 & & CH_2 & & (CH_2)_3 & & (CH_2)_4 \\
| & & | & & | & & | \\
CHCOO^- & & CH_2 & & NH_3^+ & & NH_2^+ \\
| & & | & & & & | \\
NH_3^+ & & NH_3^+ & & & & (CH_2)_3 \\
& & & & & & | \\
& & & & & & NH_3^+
\end{array}
$$

Ornithine   Putrescine   Spermidine   Spermine

Scheme I. Polyamine biosynthesis from ornithine
Reaction I is catalysed by the pyridoxal-phosphate enzyme ODC with the liberation of carbon dioxide. Reactions II and III both involve the conversion of decarboxylated S-adenosylmethionine to 5'-methylthioadenosine and are catalysed by spermidine synthase and spermine synthase, respectively.

regulation in which the unique regulatory protein antizyme plays a principal role. This is also the first demonstration that a natural enzyme is degraded *in vitro* by a specific protease in an ATP-dependent manner, and the first established case of degradation of a non-ubiquitinated protein by the 26S proteasome, which has been regarded as being specific to ubiquitinated proteins[1].

In this article, progress in the research on ODC degradation and its regulation will be reviewed briefly. For further details, see reference 10.

Polyamines destabilize ODC

The rapid turnover of ODC was first reported in 1969 by Russell and Snyder[11], who observed that ODC activity in regenerating rat liver decreased upon administration of cycloheximide, an inhibitor of protein synthesis, with a half-life as short as 11 min. The turnover rate of ODC varies depending on cellular and environmental conditions. Thus ODC induction by various stimuli is often accompanied by its several-fold stabilization, whereas the resulting increase in cellular polyamine levels later causes destabilization of ODC.

ODC in higher animals is subject to negative feedback regulation by polyamines. Exogenously added polyamines not only suppress ODC induction by various stimuli, but also elicit a rapid decay of pre-induced ODC activity. The decay of ODC activity, and ODC protein as well, by high doses of polyamines shows a characteristic time-course: after an initial lag period of less than 1 h, the enzyme activity and protein decrease at accelerating rates, which become higher than the decay rates after the addition of cycloheximide. It is important to note that polyamine-induced destabilization of ODC is inhibited by cycloheximide, but not by actinomycin D, an inhibitor of RNA synthesis, suggesting that some protein that is induced by polyamines at a post-transcriptional level plays an essential role in the polyamine-induced destabilization of ODC.

Possible role of antizyme in ODC degradation

In 1976, antizyme was discovered by Canellakis and co-workers in rat liver and some cultured cells as an ODC inhibitory protein inducible by polyamines[5]. Antizyme binds specifically to ODC, inhibiting its activity. The inactive ODC–antizyme complex can be dissociated at high ionic strength into ODC monomers (50 kDa) and antizyme (26.5 kDa, monomer) which can be separated by gel filtration. Induction of antizyme is inhibited by cycloheximide, but not by actinomycin D. Although its cellular content is very low, antizyme was purified 600000-fold to homogeneity from rat liver in 1984.

As for the metabolic fate of the ODC–antizyme complex, Canellakis and co-workers originally suggested two possibilities, namely more rapid degradation or potential storage of ODC[5]. To clarify the physiological role of antizyme, simple and reliable methods for assaying the ODC–antizyme complex were developed. The amount of the inactive complex is measured as the amount of active ODC released by excess of either difluoromethylornithine (DFMO)-inactivated ODC or antizyme inhibitor: DFMO is a mechanism-based irreversible inhibitor of ODC; antizyme inhibitor is a protein that binds with antizyme with higher affinity than ODC and, therefore, replaces ODC in the complex[12,13].

Analyses of cellular levels of the ODC–antizyme complex in rat liver and hepatoma tissue culture (HPC) cells revealed that although the amount of the complex increased temporarily on polyamine treatment, it was too small to account for the accelerated decrease in ODC activity. In the absence of exogenous polyamines, the cellular level of the complex is much lower than the induced levels of ODC, and during ODC induction upon change of medium the complex increases to a peak several hours after the ODC peak. Importantly, a good correlation was found between the antizyme/ODC ratio and the reciprocal of the half-life of ODC after addition of cycloheximide, suggesting that binding with antizyme is the rate-limiting step in ODC degradation[14]. In HMOA cells — ODC-stabilized variant cells derived from HTC cells, selected for their resistance to α-methylornithine — ODC–antizyme complex was markedly accumulated, especially upon change of medium, especially after the ODC peak[13]. Based on these results, together with the fact that both ODC destabilization and antizyme induction are elicited by polyamines without RNA synthesis, a working hypothesis was presented to show that antizyme accelerates ODC degradation in a recycling manner[13,14].

Forced expression of antizyme destabilizes ODC

Definite proof of the essential role of antizyme in ODC degradation was obtained by examining the effect of expression of transfected antizyme

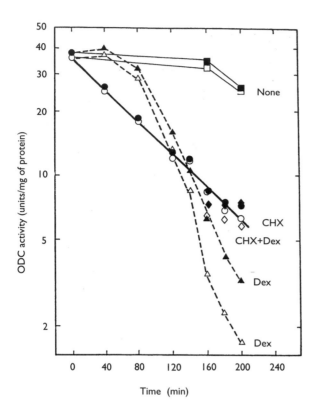

Figure 1. Effect of dexamethasone-induced antizyme on ODC decay in HZ7 cells

Dexamethasone (Dex, 1mM) and/or cyclohex-imide (CHX, 50 μg/ml) were added 2.5 h after change of the medium. ODC activity (open symbols) and total ODC (free ODC plus ODC–antizyme complex; closed symbols) were determined in the absence and presence of antizyme inhibitor, respectively. Symbols: □,■, no addition; ○,●, with CHX; ▲,△, with Dex; ◇,◆, with CHX plus Dex. Modified from reference 7.

cDNA[7]. The cDNA, Z1, which lacks the 5′-untranslated region and a small adjacent part of the coding region, was cloned from a rat liver cDNA library using monoclonal anti-antizyme antibody as a probe. In HZ7 cells (HTC cells stably transfected with antizyme cDNA Z1, constructed downstream of a glucocorticoid-inducible promoter), dexamethasone, a synthetic glucocorticoid, induced active antizyme in the absence of exogenous polyamines and elicited accelerating decay of ODC, which became more rapid than ODC decay in the presence of cycloheximide (Figure 1). The ODC destabilization was inhibited by cycloheximide. These results are essentially the same as those obtained with exogenous polyamines and clearly indicate that antizyme mediates the effect of polyamines in destabilizing ODC.

The role of antizyme was further confirmed by the following two discoveries: first, chimaeric mouse-trypanosome ODCs, which cannot bind with antizyme, are not destabilized by polyamines[15]; secondly, antizyme stimulates ODC degradation in cell-free systems[8,9].

ODC degradation *in vitro* requires ATP and antizyme

Analyses in a cell-free system are essential for revealing the molecular mechanism of any cellular function. Kahana and co-workers showed that

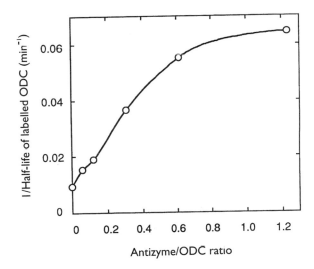

Figure 2. Dose-dependent response of ODC degradation *in vitro* to antizyme
Extracts of ODC-overproducing DF3 cells were incubated with ^{35}S-labelled ODC, ATP and various amounts of antizyme for 60 min. Labelled ODC that remained in the mixture was immunoprecipitated and separated by SDS/PAGE and its radioactivity was measured. The half-life of ODC decay was calculated from data obtained at three points in time. Modified from reference 8.

[35]S-labelled ODC synthesized in a reticulocyte lysate is degraded in such a system in an ATP-dependent but ubiquitin-independent manner[16]. These characteristics resemble those of ODC degradation in whole cells. Their *in vitro* system, however, did not degrade exogenous ODC isolated from other cells, leaving some doubt as to the physiological role of their system.

In cell-free extracts of ODC-overproducing CHO (DF3) cells and HTC cells, both endogenous and exogenous ODCs were shown to be degraded efficiently in the presence of ATP and antizyme[8,9]. Antizyme acts in a dose-dependent manner and its maximal effect is obtained when the activity ratio of antizyme to ODC approaches unity, indicating that antizyme acts by binding with ODC (Figure 2)[8]. The fact that a small amount of antizyme accelerates the degradation of much larger amounts of ODC indicates that antizyme acts in a recycling manner *in vitro* as well as *in vivo*. These results not only support the essential role of antizyme in ODC degradation, but also confirm that the system reflects ODC degradation in whole cells. The reticulocyte lysate system developed by Kahana and co-workers[16] was also shown later to be antizyme dependent.

ODC is degraded by the 26S proteasome

Analyses of ODC degradation in cell-free systems led to the identification of the protease responsible for ODC degradation[9]. Three major intracellular proteases are known, namely lysosomal cathepsins, cytosolic calpains and the

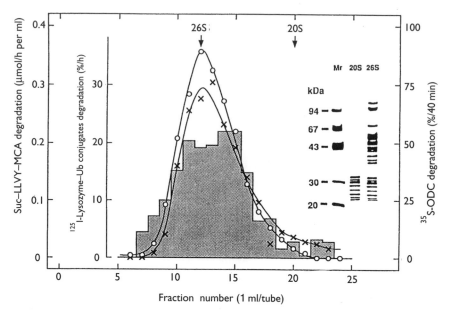

Figure 3. Distribution of ODC-degrading activity and 26S proteasome activities on glycerol density-gradient centrifugation of 26S proteasomes purified from rat liver
ODC-degrading activity (shaded histograms) was measured with ^{35}S-labelled ODC. Proteasome activity was measured with ^{125}I-labelled lysozyme–ubiquitin conjugate (O) or fluorogenic peptides (×) as substrate. The inset shows SDS/PAGE of 20S and 26S proteasomes. Reproduced from reference 9 with permission.

multicatalytic protease complex (proteasomes). The effects of various inhibitors suggested that the proteasome is a likely candidate. Removal of proteasomes from HTC cells by immunoprecipitation with anti-proteasome antibody resulted in parallel dose-dependent decreases in both ODC-degrading activity and proteasome activity, and almost complete loss of both activities was attained with a high concentration of the antibody. This indicates that proteasomes are the proteases that degrade ODC. Proteasomes are known to exist in two types: the 20S proteasome (700 kDa) consisting of 13–15 subunits, each of molecular mass 21–31 kDa, and the 26S proteasome (2000 kDa) which is made up of a 20S proteasome core and about 15 regulatory subunits, each of 35–110 kDa (Figure 3, inset)[17]. The 26S proteasome catalyses ATP-dependent degradation of polyubiquitinated proteins. Upon fractionation of purified 26S proteasome by glycerol density-gradient centrifugation, the profile of ODC-degrading activity coincided well with the profiles of proteasome activities (Figure 3). Degradation of ODC by the purified 26S proteasome, which contains neither ubiquitin nor ubiquitin-conjugating enzymes, requires ATP hydrolysis and is antizyme-dependent. These results clearly indicate that ODC is degraded by the 26S proteasome in an ATP- and antizyme-dependent manner, without ubiquitination (Figure 4).

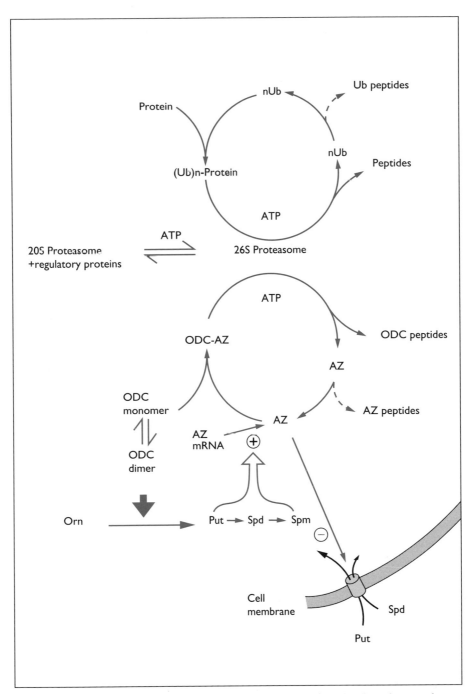

Figure 4. Diagram of negative-feedback regulation of ODC and polyamine uptake mediated by antizyme

Abbreviations used: AZ, antizyme; Ub, ubiquitin; Orn, ornithine; Put, putrescine; Spd, spermidine; Spm, spermine.

Functional regions of ODC and antizyme for ODC degradation

Several lines of evidence indicate that the C-terminal region of mouse ODC plays an essential role in its rapid degradation (Figure 5). This region is absent in trypanosomal ODC, which is stable, and mouse ODC is stabilized by deletion of 37 C-terminal amino acids[18]. Fusion of the C-terminal region of mouse ODC to either end of dihydrofolate reductase confers instability to the stable enzyme *in vitro*. Even a single amino acid replacement in this region (Cys[441] to Trp[441]) resulted in ODC stabilization in HMOA cells. The C-terminal region contains one of two PEST sequences of ODC, which are rich in proline (P), glutamic acid (E), serine (S) and threonine (T) and have been postulated by Rechsteiner and co-workers to play an important role in the selective degradation of short-lived proteins[19]. The PEST sequence appears to be related to rapid proteolysis of ODC, but not to be sufficient alone as a signal, since truncation of five C-terminal amino acids which are outside the PEST sequence also stabilizes ODC; the other PEST region is dispensable and so has no role in ODC degradation.

Trypanosome ODC does not interact with rat antizyme. Examination of various trypanosome/mouse chimaeric or deletion mutants of ODC revealed that an internal region of mouse ODC (amino acids 117 to 140) is both necessary and sufficient for antizyme binding and inhibition of ODC activity, but not for polyamine-dependent ODC destabilization, which also requires the C-terminal region[15]. These results indicated that antizyme binding is needed for polyamine-dependent degradation of ODC and that both the constitutive (polyamine-independent) and polyamine-dependent degradation of ODC need the C-terminal region. It is likely that the C-terminal region is recognized and attacked by the 26S proteasome, whereas antizyme binding induces a conformational change in the ODC molecule, resulting in exposure of the C-terminal region. Similar surveys of functional regions of antizyme indicated that although the C-terminal half of antizyme is sufficient for binding with ODC and inhibition of its activity, an adjacent region is also required for destabilization of ODC.

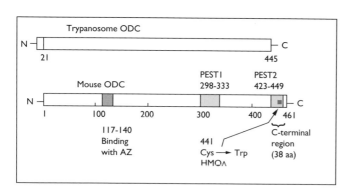

Figure 5. Diagram of mouse and trypanosome ODCs showing location of functional regions and PEST sequences Abbreviations used: aa, amino acids; AZ, antizyme.

Current model of antizyme-dependent degradation of ODC

Mouse ODC is a homodimer of a subunit with 461 amino acid residues. Two active sites are located at the interface between the subunits, each consisting of amino acids from both subunits. This is consistent with previous observations that the active ODC dimer is dissociated into inactive monomers at high ionic strength. The two subunits dissociate and reassociate unusually rapidly, so that randomization is complete within 5min under physiological conditions. Antizyme binds to ODC monomer[20], preventing its reconversion to the dimeric state and eliciting a conformational change to expose the C-terminal region, which is recognized and attacked by the 26S proteasome. Consistent with this model, ODC subunit recognition by the proteolytic machinery occurs only in the *cis* conformation, not in the *trans* conformation. Thus, when degradation of an ODC heterodimer composed of one unstable wild-type subunit and one stable mutant subunit was tested in a reticulocyte lysate, each subunit maintained its own degradation characteristics and was not affected by its partner subunit. Also consistent with this model, ODC mutants in which well-conserved glycine-387 was replaced by any other amino acid (except alanine) were unable to form a stable dimer and were short-lived.

ODC degradation need not be associated with antizyme degradation, since maltose-binding protein–antizyme fusion protein, which is quite stable, can accelerate ODC degradation *in vitro*. Antizyme has a short half-life comparable with that of ODC *in vivo*[5], and, therefore, the antizyme/ODC ratio may remain unchanged during ODC decay in the presence of cycloheximide. This may be the reason why ODC activity decays with first-order kinetics (fixed degradation rate) upon cycloheximide treatment.

Supplements

Recent studies have revealed that antizyme also mediates, either directly or indirectly, the rapid feedback repression of polyamine uptake by polyamines[21]. Therefore, antizyme is a bifunctional regulatory protein, which efficiently prevents excess accumulation of cellular polyamines (Figure 4).

Polyamines induce antizyme at the translational level by causing a programmed, ribosomal frameshift which has not previously been observed in the expression of eukaryote genes[22].

Perspectives

Twenty-five years have passed since the first report of the surprisingly rapid turnover of ODC. Major aspects of the molecular mechanism of polyamine-regulated ODC degradation have been clarified by studies with gene technology in the last five years. The ODC degradation pathway is unique in several features, such as the key role played by antizyme, frameshift-based induction of antizyme, and the proteolysis by the 26S proteasome without ubiquitination. Degradation pathways of more and more short-lived proteins will be clarified in the near future. Recent results strongly suggest that oncoproteins such as c-Myc, tumour suppressors such as p53, and cell-cycle regulators such as cyclins are degraded by 26S proteasome only after ubiquitination. The degradation of key enzymes of metabolism other than ODC has not been studied in detail at the molecular level. Several key enzymes, such as glutamine synthetase (EC 6.3.1.2) and hydroxymethyl glutaryl-CoA reductase (EC 1.1.1.88), are known to be destabilized by their products. The degradations of some may be by non-ubiquitinated pathways and may involve antizyme-like regulatory proteins.

Many problems concerning ODC degradation remain to be clarified. These include the role of antizyme inhibitor, the mechanism of ODC stabilization caused by hypotonicity or amino acids, the mechanism of polyamine-induced frameshifting, and the reason for the ATP requirement for ODC degradation. So far, the antizyme-dependent ODC degradation pathway has been observed only in mammals, birds and frogs. Three antizymes were found in *Escherichia coli* by Canellakis and co-workers, but their physiological roles appear not to be related with ODC degradation, since ODC is stable in this bacterium. It would be interesting to know how and when the antizyme-dependent ODC degradation pathway appeared during evolution.

I would like to thank Y. Murakami, S. Matsufuji and others for their great contribution in propelling the study described in this article. I am also grateful to K. Tanaka and others of Tokushima University for the fascinating collaborative study on the role of proteasomes in ODC degradation.

References

1.   Hershko, A. & Ciechanover, A. (1992) The ubiquitin system for protein degradation. *Annu. Rev. Biochem.* **61**, 761–807

2.   Pegg, A.E. (1986) Recent advances in the biochemistry of polyamines in eukaryotes. *Biochem. J.* **234**, 249–262

3.   Hayashi, S. (1989) Multiple mechanisms for the regulation of mammalian ornithine decarboxylase. In *Ornithine Decarboxylase: Biology, Enzymology, and Molecular Genetics* (Hayashi, S., ed.), pp. 35–45, Pergamon Press, New York

4.   Morris, D.R. (1991) A new perspective on ornithine decarboxylase regulation: prevention of polyamine toxicity is the overriding theme. *J. Cell. Biochem.* **45**, 1–4

5.   Heller, J.S., Fong, W.F. & Canellakis, E.S. (1976) Induction of a protein inhibitor to ornithine decarboxylase by the end products of its reaction. *Proc. Natl. Acad. Sci. U.S.A.* **73**, 1858–1862

6.   Hayashi, S. & Canellakis, E.S. (1989) Ornithine decarboxylase antizymes. In *Ornithine Decarboxylase: Biology, Enzymology, and Molecular Genetics* (Hayashi, S., ed.), pp. 47–58, Pergamon Press, New York

7.   Murakami, Y., Matsufuji, S., Miyazaki, Y. & Hayashi, S. (1992) Destabilization of ornithine decarboxylase by transfected antizyme gene expression in hepatoma tissue culture cells. *J. Biol. Chem.* **267**, 13138–13141

8.   Murakami, Y., Tanaka, K., Matsufuji, S., Miyazaki, Y. & Hayashi, S. (1992) Antizyme, a protein induced by polyamines, accelerates the degradation of ornithine decarboxylase in Chinese-hamster ovary-cell extracts. *Biochem. J.* **283**, 661–664

9.   Murakami, Y., Matsufuji, S., Kameji, T. *et al.* (1992) Ornithine decarboxylase is degraded by the 26S proteasome without ubiquitination. *Nature (London)* **360**, 597–599

10. Hayashi, S. & Murakami, Y. (1995) Rapid and regulated degradation of ornithine decarboxylase. *Biochem. J.*, in the press

11. Russell, D.H. & Snyder, S.H. (1969) Amine synthesis in regenerating rat liver: extremely rapid turnover of ornithine decarboxylase. *Mol. Pharmacol.* **5**, 253–262

12. Fujita, K., Murakami, Y. & Hayashi, S. (1982) A macromolecular inhibitor of the antizyme to ornithine decarboxylase. *Biochem. J.* **204**, 647–652

13. Murakami, Y., Fujita, K., Kameji, T. & Hayashi, S. (1985) Accumulation of ornithine decarboxy-lase–antizyme complex in HMOA cells. *Biochem. J.* **225**, 689–697

14. Murakami, Y. & Hayashi, S. (1985) Role of antizyme in degradation of ornithine decarboxylase in HTC cells. *Biochem. J.* **226**, 893–896

15. Li, X. & Coffino, P. (1992) Regulated degradation of ornithine decarboxylase requires interaction with the polyamine-inducible protein antizyme. *Mol. Cell. Biol.* **12**, 3556–3562

16. Bercovich, Z., Rosenberg-Hasson, Y., Ciechanover, A. & Kahana, C. (1989) Degradation of ornithine decarboxylase in reticulocyte lysate is ATP-dependent but ubiquitin-independent. *J. Biol. Chem.* **264**, 15949–15952

17. Tanaka, K., Tamura, T., Yoshimura, T. & Ichihara, A. (1992) Proteasomes: protein and gene structures. *New Biol.* **4**, 173–187

18. Ghoda, L., van Daalen Wetters, T., Macrae, M., Ascherman, D. & Coffino, P. (1989) Prevention of rapid intracellular degradation of ODC by a carboxyl-terminal truncation. *Science* **243**, 1493–1495

19. Rogers, S., Wells, R. & Rechsteiner, M. (1986) Amino acid sequences common to rapidly degraded proteins: the PEST hypothesis. *Science* **234**, 364–368

20. Mitchell, J.L.A. & Chen, H.J. (1990) Conformational changes in ornithine decarboxylase enable recognition by antizyme. *Biochim. Biophys. Acta* **1037**, 115–121

21. Mitchell, J.L.A., Judd, G.G., Bareyal-Leyser, A. & Ling, S.Y. (1994) Feedback repression of polyamine transport is mediated by antizyme in mammalian tissue-culture cells. *Biochem. J.* **299**, 19–22

22. Matsufuji, S., Matsufuji, T., Miyazaki, Y., Murakami, Y., Atkins, J.F., Gesteland, R.F. & Hayashi, S. (1995) Autoregulatory frameshifting in decoding mammalian ornithine decarboxylase antizyme. *Cell*, in the press

4

The chloroplast genome

Masahiro Sugiura

Center for Gene Research, Nagoya University, Nagoya 464-01, Japan

Introduction

Chloroplasts are the green organelles in plant leaves that contain the entire machinery necessary for the process of photosynthesis. Chloroplasts also participate in the biosynthesis of amino acids, nucleotides, lipids and starch. At the beginning of this century, Baur and Correns independently found non-Mendelian (maternal) inheritance based on studies of variegation in higher plants. Further analysis of variegation revealed that the genetic determinants for these characters were associated with chloroplasts.

Half a century later, the demonstration of a unique DNA species in chloroplasts[1] has led to intensive studies of both the structure and expression of chloroplast genomes. In addition, as chloroplast DNA (ctDNA) molecules are relatively small and simple, they were selected as one of the first targets of 'the genome projects'. The entire nucleotide sequences of eight chloroplast genomes have been determined to date, disclosing an enormous amount of functional and evolutionary information.

Structure of ctDNA

Almost all ctDNAs are circular molecules ranging in size from 120 kb to 160 kb[2]. The siphonous green alga *Codium fragile* has the smallest ctDNA known (89 kb), while the green alga *Chlamydomonas moewusii* has the largest (292 kb). Non-photosynthetic plants generally contain much-reduced chloroplast genomes of 50–73 kb due to the loss of many normal chloroplast genes. Entire ctDNA sequences are now available for *Nicotiana tabacum* (tobacco)[3], *Marchantia polymorpha* (liverwort)[4], *Oryza sativa* (rice)[5], *Epifagus virginiana* (beechdrops)[6], *Euglena gracilis* (a green alga)[7], *Pinus thunbergii* (black pine)[8],

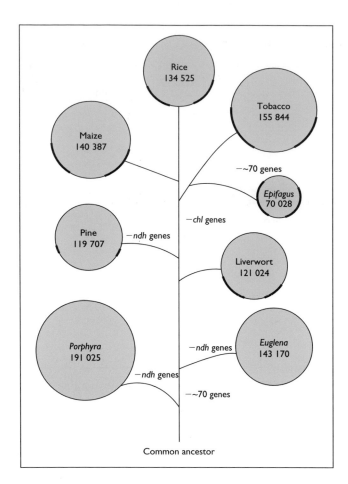

Figure 1. A phylogenetic representation of the entirely sequenced chloroplast genomes
The entire length is shown (kb). Minus indicates gene losses. Bold lines denote IRs.

Zea mays (maize; H. Kössel, personal communication) and *Porphyra purpurea* (a red alga[9]; M. Reith, personal communication), as shown in Figure 1.

One of the outstanding features of the ctDNAs is the presence of a large inverted repeat (IR) which ranges from 5 kb to 76 kb in length[2]. This arrangement results in doubling of the rRNA genes as well as other genes included within the IRs: only rRNA genes in the 5 kb IR of brown algae, 10 additional genes in the 25 kb IR of tobacco and over 40 additional genes in the 76 kb IR of geranium. The ctDNAs of some legumes, conifers and algae are exceptions to this pattern and lack IRs. It has been suggested that the IRs were present in the common ancestor of land plants and one segment of the IR was lost in some legumes and conifers during evolution[2]. However, the loss of the IR is partial, at least in black pine, as its genome retains a remnant of the large IR[8].

Genes on ctDNA

Most ctDNAs are now known to contain 30–36 RNA genes and over 60 genes encoding proteins[10]. These genes can be categorized into two main classes: genes involved in transcription/translation and those related to photosynthesis

Table I. Number of identified genes on the entirely sequenced chloroplast genomes

	Land plants		Algae	
	Photosynthetic	Epifagus	Euglena	Porphyra
Total	101–107	40	82	182
Genetic system				
rRNA	4	4	3	3
tRNA	30–32	17	27	35
r-protein	2–21	15	21	46
Other	5–6	2	4	18
Photosynthesis				
Rubisco and thylakoid	29–30	0	26	40
ndh	11	0	0	0
Biosynthesis and miscellaneous	1–5	2	1	40
Number of introns	18–21	6	149	0

(Table 1). The gene order present in tobacco is most representative of land plants, probably reflecting the ancestral gene order among higher plants (Figure 2).

It was originally believed that the gene organization of ctDNAs was relatively uniform from species to species. However, recent analysis of chloroplast genomes from a variety of algae has revealed that this is not always the case. New genes have been found one after another in chloroplast genomes from algae. The *Porphyra* chloroplast genome contains over 70 new genes not found in the seven other sequenced genomes; one-third of these belong to an additional class of chloroplast genes encoding protein involved in the biosynthesis of amino acids, fatty acids, pigments and so on[9]. The *Porphyra* chloroplast genome appears to be the most 'primitive' chloroplast genome described to date.

ctDNA encodes all rRNAs and 27–35 tRNA species. All 61 possible codons are used in chloroplast genes encoding polypeptides. The minimum number of tRNA species required for translation of all 61 codons is 32, if normal wobble base-pairing occurs in codon–anticodon recognition. No tRNA which recognizes several codons according to normal wobble base-pairing has been found. If the 'two-out-of-three' and 'U:N wobble' mechanisms operate in the chloroplast, chloroplast-encoded tRNAs are probably sufficient to read all 61 codons[3]. The chloroplast genome of beechdrops lacks 13 of the tRNA genes that are found in tobacco and it seems probable that nuclear-encoded tRNAs are imported into the chloroplast to effect translation[6]. However, no direct evidence is provided to support RNA

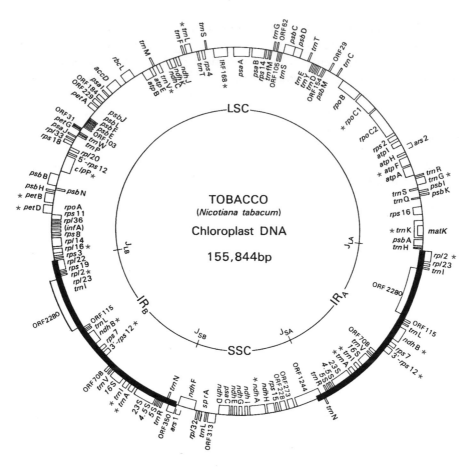

Figure 2. Gene map of the tobacco chloroplast genome
Genes shown inside the circle are transcribed clockwise, and genes on the outside are transcribed anticlockwise. Asterisks denote split genes. Abbreviation used: IRF, intron-containing reading frame. Bold lines show IRs. Reproduced from the Research Grant Progress Report (1989) with minor revisions.

transport into chloroplasts. In addition to rRNAs and tRNAs, novel RNA species encoded by ctDNA have recently been found: *Chlamydomonas tsc*A RNA involved in *trans*-splicing (see below); tobacco *spr*A RNA for rRNA maturation; and *Porphyra* RNaseP RNA for tRNA processing.

Sixty or more polypeptides are encoded by ctDNA. These include transcription and translation components, such as RNA polymerase subunits, ribosomal proteins and translation factors, and components of the photosynthesis apparatus, such as ribulose bisphosphate carboxylase subunits and thylakoid membrane subunits. However, chloroplast apparatuses are composed not only of chloroplast-encoded subunits but also of many nuclear-encoded proteins. For example, two-thirds of ribosomal proteins are imported from the nucleo-cytoplasm.

Determination of ctDNA sequences revealed the existence of an unexpected set of genes, *ndh* genes, whose predicted amino acid sequences resemble the components of the respiratory-chain NADH dehydrogenase from mitochondria[3,4]. This observation suggests the existence of a respiratory chain in chloroplasts, although it remains to be determined whether or not an active NADH dehydrogenase is present in land-plant chloroplasts. The *ndh* genes are absent in *Euglena* and *Porphyra* chloroplasts. More interestingly, the chloroplast genome of black pine contains no functional *ndh* genes, although several *ndh* pseudogenes are present. These observations suggest the transfer of all *ndh* genes into the nuclear genome in some higher plants and algae.

Chloroplast gene expression

Chloroplast genes are transcribed by more than one class of DNA-dependent RNA polymerase[11]. Sequence analysis of chloroplast DNAs revealed that these DNAs contain homologues (*rpoA, rpoB, rpoC*) of the genes encoding subunits α, β and β′ of *Escherichia coli* RNA polymerase, indicating the presence of an *E. coli*-like RNA polymerase in chloroplasts. However, RNA synthesis was detected in chloroplasts lacking functional chloroplast ribosomes and, consequently, chloroplast translation products including the subunits of *E. coli*-like RNA polymerase, which implies the existence of a second RNA polymerase, encoded entirely by the nuclear genome and transported into chloroplasts. In addition, an RNA polymerase tightly bound to DNA and transcribing rRNA genes and a T7-like single-subunit RNA polymerase were also reported in chloroplasts.

The upstream regions of many transcriptional initiation sites from chloroplast genes contain sequence motifs similar to the '−10' and '−35' *E. coli* promoter elements, which are likely to be recognized by *E. coli*-like RNA polymerase. However, a class of chloroplast tRNA genes has been identified which does not require the 5′ upstream regions for transcription. The *psbA* genes in higher plants contain both −10/−35 motifs and between them the eukaryotic TATA box. Multiple transcription initiation sites are often observed in many chloroplast operons, including rRNA genes and *atpB/E* genes, some of which lack any known promoter elements, suggesting the existence of an additional novel type of promoter. Thus the chloroplast genome contains at least three structurally distinct promoters, the situation of which is compatible with the existence of multiple RNA polymerases. Chloroplast genes are generally co-transcribed, and polycistronic transcripts are processed into many overlapping shorter RNA species. Post-transcriptional processing of primary transcripts in chloroplasts consists of multiple reactions, including cutting, 3′-trimming, *cis/trans*-splicing and RNA editing, and represents a critical step in the control of chloroplast gene expression[12]. Processing of rRNA and tRNA precursors is also required to form functional RNA species. A small chloroplast-encoded RNA of 218 nt (*sprA* RNA) has been found in tobacco chloroplasts and shows potential base-pairing with the leader sequence of pre-16S rRNA, suggesting a role for the RNA in 16S rRNA maturation[13].

trans-splicing

Some of the chloroplast genes contain introns. Intron content is conserved among land plants, whereas that in algae is highly diverged: 149 introns in *Euglena* account for about 40% of the genome, whereas there are none in *Porphyra* (Table 1). Most genes possessing introns in higher plants contain single introns, while many polypeptide-encoding genes have multiple introns in *Euglena* and *Chlamydomonas*. Six chloroplast tRNA genes in higher plants have introns, but none are known in algal chloroplasts.

Among all the chloroplast genes containing introns, the ribosomal protein S12 genes (*rps12*) in land plants and the gene encoding photosystem I apoprotein A1 (*psaA*) in *Chlamydomonas reinhardtii* have the most striking features. The tobacco *rps12* gene is divided into one copy of exon 1 (38 codons) and two copies of exon 2 (78 codons)/intron (536 bp)/exon 3 (7 codons)[3]. The two portions are widely separated and are transcribed independently. These two transcripts are spliced in *trans* to produce a mature mRNA for S12 (Figure 3). The 3′-flanking sequence of exon 1 and the 5′-flanking sequence of exon 2 fit the conserved boundary sequences of chloroplast

Figure 3. Process for *rps12* and *psaA* mRNA maturation from separate pre-mRNAs by *trans*-splicing
Upper circles show the location of gene portions and the direction of transcription. Bold lines are IR. Tobacco *rps12* is divided into exon 1 (E1) and exon 2/intron/exon 3 (E2+3). *Chlamydomonas psaA* consists of exons 1 (E1), exon 2 (E2) and exon 3 (E3). The splicing pathway is shown in the lower part. Boxes indicate exons and a wavy line denotes an intron.

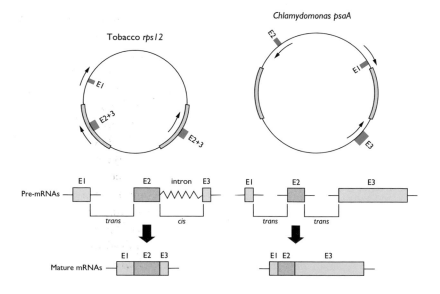

introns. It is noteworthy that tobacco *rps12* requires both *cis*- and *trans*-splicing to produce the mature mRNA.

The *Chlamydomonas psaA* gene is also divided into three parts[14] (Figure 3). The first exon of 30 codons is 50 kb away from the second exon (60 codons), which is itself 90 kb away from exon 3 (661 codons). All exons are flanked by the consensus intron boundary sequences. The three exons are transcribed independently as precursors, and the synthesis of mature *psaA* mRNA involves *trans* assembly of these three separate transcripts. At least one additional chloroplast locus *(tscA)* encoding a 450 base RNA is required for *trans*-splicing of exons 1 and 2. A model has been proposed in which the *tscA* RNA bridges the 3´-flanking sequence of exon 1 and the 5´-flanking sequence of exon 2[14].

RNA editing

Another interesting reaction recently shown to occur during the chloroplast post-transcriptional processing is RNA editing. RNA editing is defined as the post-transcriptional modification of pre-RNA to alter its nucleotide sequence through the insertion and deletion of nucleotides, or specific nucleotide substitutions, so as to yield functional RNA species.

The genes for ribosomal protein L2 (*rpl2*) from maize and rice are known to have an ACG codon at the position corresponding to an ATG initiation codon in other plant *rpl2*s analysed. Analysis of the maize *rpl2* mRNA revealed that the ACG codon is converted to an AUG initiation codon by a C to U base change[15] (Figure 4). RNA editing is not limited to initiation codons but has been observed at internal codons. All the edited codons restored amino acids that are conserved in the corresponding proteins of other plants, and, therefore, RNA editing was believed to be functionally significant. However, recent analysis of transcripts from black pine chloroplasts has shown that one of the C to U base changes occurs at the third position of a codon and no amino acid substitution is expected (silent editing) (T. Wakasugi, T. Hirose and M. Sugiura, unpublished work). Moreover, two C to U alterations were

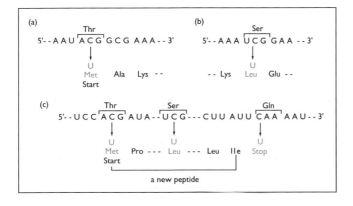

Figure 4. RNA editing in chloroplasts
(a) Creation of an initiation codon.
(b) Internal editing restores a codon for a conserved amino acid.
(c) Creation of both initiation and stop codons to yield a new reading frame.

observed to produce both start and stop codons at once within a transcript from black pine chloroplasts. This editing process creates a new polypeptide gene which cannot be deduced merely from the DNA sequence. To date, 20–30 editing sites have been observed in the chloroplast transcripts of monocots, dicots or gymnosperms, while none has been reported in those of liverwort and algae.

The existence of *trans*-splicing and RNA editing steps indicates that chloroplast protein sequences cannot always be predicted from continuous ctDNA sequences, and, therefore, post-transcriptional processing is essential for accurate chloroplast gene expression. The question of why chloroplasts have such elaborate processes remains unanswered. It has been postulated that these steps are remnants of the old RNA world to transfer proper genetic messages into proteins, and that present day nuclear genomes have acquired accurate genetic information in their DNA sequences, co-linear with peptide sequences, over the course of evolution. Thus the chloroplast genetic system may offer a window on evolutionary antiquity.

Summary

- *The chloroplast genome consists of homogeneous circular DNA molecules. To date, the entire nucleotide sequences (120–190 kbp) of chloroplast genomes have been determined from eight plant species.*
- *The chloroplast genomes of land plants and green algae contain about 110 different genes, which can be classified into two main groups: genes involved in gene expression and those related to photosynthesis.*
- *The red alga* Porphyra *chloroplast genome has 70 additional genes, one-third of which are related to biosynthesis of amino acids and other low molecular mass compounds.*
- *Chloroplast genes contain at least three structurally distinct promoters and transcribe two or more classes of RNA polymerase.*
- *Two chloroplast genes,* rps12 *of land plants and* psaA *of* Chlamydomonas, *are divided into two to three pieces and scattered over the genome. Each portion is transcribed separately, and two to three separate transcripts are joined together to yield a functional mRNA by* trans-*splicing.*
- *RNA editing (C to U base changes) occurs in some of the chloroplast transcripts. Most edited codons are functionally significant, creating start and stop codons and changing codons to retain conserved amino acids.*

I thank Dr T. Wakasugi for discussion, Dr H. Kössel and Dr M. Reith for unpublished data, and Dr R. Whittier for critical reading of the manuscript.

References

1. Sager, R. & Ishida, M.R. (1963) Chloroplast DNA in *Chlamydomonas*. *Proc. Natl. Acad. Sci. U.S.A.* **50**, 725–730

2. Palmer, J.D. (1991) Plastid chromosomes: structure and evolution. In *The Molecular Biology of Plastids* (Bogorad, L. & Vasil, I.K., eds.), pp. 5–53, Academic Press, San Diego

3. Shinozaki, K., Ohme, M., Tanaka, M. *et al.* (1986) The complete nucleotide sequence of the tobacco chloroplast genome: its gene organization and expression. *EMBO J.* **5**, 2043–2049

4. Ohyama, K., Fukuzawa, H., Kohchi, T. *et al.* (1986) Chloroplast gene organization deduced from complete sequence of liverwort *Marchantia polymorpha* chloroplast DNA. *Nature (London)* **322**, 572–574

5. Hiratsuka, J., Shimada, H., Whittier, R. *et al.* (1989) The complete sequence of the rice (*Oryza sativa*) chloroplast genome: intermolecular recombination between distinct tRNA genes accounts for a major plastid DNA inversion during the evolution of the cereals. *Mol. Gen. Genet.* **217**, 185–194

6. Wolfe, K.H., Morden, C.W. & Palmer, J.D. (1992) Function and evolution of a minimal plastid genome from a nonphotosynthetic parasitic plant. *Proc. Natl. Acad. Sci. U.S.A.* **89**, 10648–10652

7. Hallick, R.B., Hong, L., Drager, R.G. *et al.* (1993) Complete sequence of *Euglena gracilis* chloroplast DNA. *Nucleic Acids Res.* **21**, 3537–3544

8. Wakasugi, T., Tsudzuki, J., Ito, S., Nakashima, K., Tsudzuki, T. & Sugiura, M. (1994) Loss of all *ndh* genes as determined by sequencing the entire chloroplast genome of the black pine *Pinus thunbergii*. *Proc. Natl. Acad. Sci. U.S.A.* **91**, 9794–9798

9. Reith, M. & Munholland, J. (1993) A high-resolution gene map of the chloroplast genome of the red alga *Porphyra purpurea*. *Plant Cell* **5**, 465–475

10. Sugiura, M. (1992) The chloroplast genome. *Plant Mol. Biol.* **19**, 149–168

11. Pfannschmidt, T. & Link, G. (1994) Separation of two classes of plastid DNA-dependent RNA polymerases that are differentially expressed in mustard (*Sinapis alba* L.) seedlings. *Plant Mol. Biol.* **25**, 69–81

12. Deng, X.-W. & Gruissem, W. (1987) Control of plastid gene expression during development: the limited role of transcriptional regulation. *Cell* **49**, 379–387

13. Vera, A. & Sugiura, M. (1994) A novel RNA gene in the tobacco plastid genome: its possible role in the maturation of 16S rRNA. *EMBO J.* **9**, 2211–2217

14. Rochaix, J.-D. (1992) Post-transcriptional steps in the expression of chloroplast genes. *Annu. Rev. Cell Biol.* **8**, 1–28

15. Hoch, B., Maier, R.M., Appel, K., Igloi, G.L. & Kössel, H. (1991) Editing of a chloroplast mRNA by creation of an initiation codon. *Nature (London)* **353**, 178–180

5

Lectins — proteins with a sweet tooth: functions in cell recognition

Nathan Sharon and Halina Lis

Department of Membrane Research and Biophysics, The Weizmann Institute of Science, Rehovot 76100, Israel

Introduction and overview

Towards the end of last century, scattered reports started to appear in the scientific literature on the occurrence in plants of proteins that possess the remarkable ability to agglutinate erythrocytes. It took nearly 50 years before it was demonstrated that concanavalin A, the first purified plant haemagglutinin, is carbohydrate specific and that it binds non-covalently to sugars on cell surfaces. In the 1940s it was found that certain haemagglutinins distinguish sharply between erythrocytes of different blood types. For instance, the agglutinin from lima bean, specific for *N*-acetylgalactosamine, agglutinates blood type A (and type AB) cells, while that from the asparagus pea is specific for fucose and agglutinates only type O cells. This finding prompted W.C. Boyd, in 1954, to coin the term lectins (from the latin *legere*, meaning to select) to emphasize the fact that these proteins are selective in their interactions with carbohydrates and cells. It also led to the conclusion that sugars are the immunodeterminants of blood type ABO specificity.

For nearly 100 years, research in this area focused on plant lectins[1,2]. During this period, hundreds of lectins have been purified, mainly by affinity chromatography on immobilized carbohydrates, and studied with respect to their chemical and biological properties. Moreover, a wide range of applications of these proteins was elaborated. Only during the last two decades has it become clear that lectins are ubiquitous in nature and that they come in a

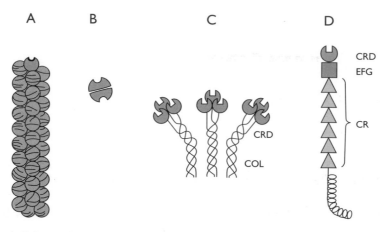

Figure 1. Schematic representation of the overall structure of lectins from different sources

(A) Part of fimbria (e.g 'type 1' and 'type P') of *Escherichia coli*. The fimbriae are an assembly of different types of subunit, only one of which (blue) contains a carbohydrate-binding site. (B) Dimeric legume lectin (e.g. from peas or from seeds of *Erythrina corallodendron*). (C) Mammalian soluble mannose-binding protein, a collectin. (D) E-Selectin; the coil represents the membrane-spanning and cytoplasmic domains. Abbreviations used: COL, collagenous region; CR, complement regulatory repeats; EGF, epidermal growth factor-like domain; CRD, carbohydrate recognition domain. The indentations represent binding sites.

variety of sizes and shapes (Figure 1). Nonetheless, many of them can be grouped into distinct families that share similarities in amino acid sequences and structural properties. In fact, sequence similarities with those of known lectins, provide a new guideline for the detection and identification of new ones. These families are often along taxonomic lines (e.g. legume and cereal lectins), although sometimes common features are found across phylogenetic barriers as well. An extreme example is *Bordetella pertussis* toxin, a bacterial lectin with peptide stretches similar to a plant and an animal lectin arranged in tandem[3].

In the past, lectins attracted interest mainly because they proved to be invaluable tools in the structural and functional investigation of glycoconjugates. Owing to their ability to discriminate between the myriad of complex carbohydrate structures found in soluble glycoproteins, on the surfaces of cells and in extracellular matrices, lectins are being employed for the detection and isolation of glycoproteins and for the partial characterization of their carbohydrate moieties. They are also used for cell identification and fractionation, as well as for following changes that occur in cell-surface sugars in processes such as development, differentiation and neoplastic transformation[1]. Another useful property of many lectins is their ability to cause mitogenic stimulation — the triggering of quiescent, non-dividing lymphocytes into a state of growth and proliferation. This provides an easy and simple means to assess the immuno-competence of patients suffering from diverse diseases. It also serves in the detection of chromosomal defects, since in the stimulated cells the chromosomes are easily visualized.

At present, attention has shifted to the question of how lectins bind carbohydrates, and what their role is in nature.

Carbohydrate binding and biological recognition

Detailed understanding of how lectins bind carbohydrates requires identification of the chemical groups on those molecules that interact with each other and of the types of bond formed (hydrogen, van der Waals, hydrophobic and electrostatic). This has now become feasible, to a large extent thanks to the development of new techniques and the refinement of old ones. In particular, high-resolution X-ray crystallography of lectins in complex with their ligands allows such identification by 'visualizing' the interactions established between the protein and the carbohydrate. Further information on the contribution of individual amino acid residues to the binding activity of a lectin is obtainable by site-directed mutagenesis of their DNA. From the data available on the dozen or so lectin–carbohydrate complexes that have been studied by these techniques[4], it is clear that the combining sites of lectins also have different shapes and that diverse amino acids are involved in the formation of bonds to the carbohydrate, although they are similar in the same family. This is illustrated in Table 1, which compares the amino acids and metal ions involved in the binding of monosaccharides to plant and animal lectins.

The main function of lectins is in cell recognition, a central event in a variety of biological phenomena, such as fertilization, organ formation or immune defence[5]. The basis of this recognition is the molecular fit between pairs of complementary structures on the surfaces of the interacting cells: one carries encoded biological information and the other is capable of deciphering the code. This concept represents an extension of the lock-and-key hypothesis originally used to explain the specificity of interactions between enzymes and

Table 1. Amino acids and metal ions in the monosaccharide-combining sites of plant and animal lectins

Families	Hydrogen bonding amino acids	Aromatic residues	Metal ions	
			Identity	Role
Plants				
Legumes	Asn, Asp	+	Ca^{2+}, Mn^{2+}	Structural
Cereals	Glu, Ser, Tyr	+ +	None	
Animals				
Galectins	Arg, Asn, His, Glu	+	None	
C-type	Asn, Glu	-	Ca^{2+}	Coordinate ligand

Additional interactions occur when the lectins bind oligosaccharides.

Table 2. Lectins in cell–cell recognition

Lectin	Role	Carbohydrate determinant
Micro-organisms		
Influenza virus		
Human isolates	Infection	NeuAc(α2-6)Gal
Avian and equine isolates	Infection	NeuAc(α2-3)Gal
Amoeba	Infection	Gal/GalNAc
Bacteria		
E. coli, type I	Infection	Mannose
E. coli, type P	Infection	Gal(α1-4)Gal
Animals		
Galectins	Metastasis (?)	Gal(β1-4)GlcNAc
C-type lectins		
Mannose-binding protein	Host antimicrobial defence	Mannose
L-selectin	Lymphocyte homing	
E-selectin	Leucocyte trafficking to sites of inflammation	SiaLe^a, SiaLe^x
P-selectin	Leucocyte trafficking to sites of inflammation	SiaLe^a, SiaLe^x

Abbreviations used: SiaLe^a, sialyl-Lewis^a or NeuAc(α2-3)Gal(β1-4)[Fuc(α1-3)]GlcNAc; SiaLe^x, sialyl-Lewis^x or NeuAc(α2-3)Gal(β1-3)[Fuc(α1-4)]GlcNAc.

their substrates. The idea that carbohydrates and lectins are eminently suitable to act in cell recognition evolved with the demonstration that both types of compound are commonly present on cell surfaces, and with the realization that carbohydrates have an enormous potential for encoding biological information. This potential derives from the ability of monosaccharides to combine with each other by a variety of linkages and also to form branched oligomers and polymers. Thus they can carry per unit weight much more information than can nucleic acids and proteins, linear polymers with a single type of linkage. Lectins possess exquisite specificity, since they not only distinguish between different monosaccharides, but can also detect subtle differences in complex carbohydrate structures. Furthermore, they combine with carbohydrates rapidly and reversibly.

That lectins and carbohydrates are partners in biological recognition was proven unequivocally by discoveries made in two seemingly unrelated areas. The first was the demonstration that lectins mediate the adhesion of many pathogenic micro-organisms to host cells, a precondition for infection; the second was the finding that they control the migration of leucocytes in the body and their recruitment to sites of inflammation (Table 2). Both activities were shown to require the binding of lectins to complementary sugars on apposing cells. These discoveries led to an explosion of activity aimed at developing carbohydrate-based inhibitors of lectins to be used as anti-adhesion drugs for the prevention of infections and inflammation[6].

Microbial lectins

Lectins are common constituents of bacterial surfaces, and are also present on other classes of micro-organism, such as protozoa (e.g. amoeba[7]), as well as viruses. Influenza virus haemagglutinin, the paradigm of viral lectins, is a surface glycoprotein specific for the most common sialic acid, N-acetyl-neuraminic acid (NeuAc)[8]. The subunit of the lectin is composed of two polypeptides, covalently linked by a single disulphide bond, and it associates non-covalently to form trimers that are located on the surface of the viral membrane. The carbohydrate-binding site is located in a pocket of one of the two polypeptide chains that comprise each of the subunits, in a domain protruding from the membrane, and is composed of amino acids that are largely conserved in the numerous strains of the virus, not all of which have the same host specificity. Thus human strains bind preferentially to the disaccharide NeuAc(α2-6)Gal found predominantly on human cells. Avian and equine strains prefer the isomeric compound, NeuAc(α2-3)Gal, which differs from NeuAc(α2-6)Gal only in the linkage position between the two sugars and is more abundant in these animals. This difference in specificity pattern appears to be dictated by the replacement of a single amino acid at position 266 of the binding site — a leucine residue in α2-6-specific strains and a glutamine residue in α2-3-specific ones.

The lectin mediates the attachment of the virus to cells by combining with sialic acid-containing carbohydrates on their surface. This is followed by fusion of the viral and cellular membranes, allowing release of the viral genome into the cytoplasm and subsequent replication. Removal of sialic acid from the cell membranes by treatment with the enzyme sialidase abolishes binding and prevents infection. Furthermore, coating of sialidase-treated cells with sialic acid-containing glycolipids restores their susceptibility to infection by the virus. It is, therefore, clear that receptor analogues that will block attachment of the virus to cells might prove to be effective antiviral drugs.

Bacterial lectins are frequently in the form of fimbriae (or pili), submicro-scopic proteinaceous appendages that protrude from the surface of the cells[9] (Figure 1). The best characterized are the 'type 1' fimbriae, specific for mannose, and the 'type P' fimbriae, specific for the disaccharide galabiose [Gal(α1-4)Gal] produced by many strains of *Escherichia coli*. These fimbriae consist of an assembly of protein subunits of several different types, only one of which carries a carbohydrate-binding site; this subunit is a minor component and is usually located laterally along the fimbriae or at their tips. Certain strains of *E. coli* isolated from humans and farm animals express fimbrial lectins specific for glycoconjugates containing sialic acid. Comparison of the amino acid sequences of the carbohydrate-binding subunit of these lectins revealed the presence of a common motif, rich in basic amino acids. Site-specific mutagenesis experiments have shown that a lysine residue and an arginine residue in this region play a part in ligand binding[10].

A major role of the microbial lectins is to mediate the adhesion of the organisms to host cells, an initial stage of infection[9]. Thus mannose and methyl α-mannoside inhibited specifically infection of the urinary tract of mice and rats by different strains of type-1 fimbriated *E. coli* and *Klebsiella pneumoniae*, respectively, while sialic acid-containing glycopeptides, administered orally, protected colostrum-deprived newborn calves against lethal doses of entero-toxigenic *E. coli* K99. Similarly, NeuAc reduced considerably colonization of lung, liver and kidney by *Pseudomonas aeruginosa* injected intravenously into mice. Further, introduction of galactose into the trachea of rabbits infected with *B. pertussis* prevented colonization of the respiratory tract by the bacteria and blocked pulmonary oedema[11]. In a clinical trial in humans, patients with otitis externa (a painful swelling, with secretion, of the external auditory canal caused by *P. aeruginosa*) were treated locally with a solution of galactose, mannose and NeuAc[12]; the results were fully comparable with those obtained with local antibiotic treatment. These findings illustrate the great potential of carbohydrates in the prevention of infections caused by bacteria that express surface lectins, and raise hopes for the development of anti-adhesive drugs for human use.

Expression of the bacterial lectins during the infection process is not always harmful to the host. Thus type-1 fimbriated bacteria may adhere to phagocytic cells, often leading to ingestion and killing of the bacteria. The process has been named lectinophagocytosis[9] in analogy with opsonophago-cytosis, in which recognition between the micro-organisms and the phagocytic cells is mediated by serum constituents termed opsonins (mainly IgG antibodies and complement fragments). Another type of lectinophagocytosis is that resulting from the binding of bacteria to the mannose-specific lectin of macrophages (see later). Lectinophagocytosis may function as a defence mechanism against microbial infection *in vivo* at sites and in situations where opsonic activity is poor, for instance the renal medulla and peritoneal cavity, especially during dialysis, or in patients infected by micro-organisms prior to the development of an immune response.

Plant lectins

As mentioned already, for a long time plants were practically the only source of lectins. Still their role in nature remains an enigma, although several proposals are currently in vogue, such as the binding of symbiotic rhizobia to the roots of leguminous plants, or the protection of plants against pathogens and predatory insects. A number of families of plant lectins have been identified; the largest and best studied is that from the seeds of legumes, with over 70 members that have been characterized[4,13]. Legume lectins consist of two or four identical or very similar subunits of 25–30 kDa, each with a single, small carbohydrate-combining site and tightly bound Ca^{2+} and Mn^{2+} which are required for activity. The primary structures of some 40 of these lectins have been established by chemical or molecular genetic techniques. They exhibit remarkable homologies, with 20% of amino acid residues invariant, including several of those involved in the binding of monosaccharides and almost all the residues that coordinate the metal ions. The three-dimensional structure of eight legume lectins, most of them in complex with mono- or oligosaccharides, has been elucidated by X-ray crystallography. The subunits are in the shape of a half-dome, with the carbohydrate-binding site forming a shallow depression at its apex. The structures are nearly superimposable, irrespective of the specificity of the lectins. The metal ions are located close to the combining site, where they help to position amino acid residues for carbo-hydrate binding but do not bind directly to the carbohydrate. In each of the legume lectins, 4–10 hydrogen bonds, some of them bridged by water molecules, as well as a few hydrophobic interactions, hold the monosaccharide in the combining site. A single hydroxyl group of the sugar may be linked to the protein by more than one hydrogen bond.

The fact that several of the amino acids that are involved in the binding of the carbohydrate (and the metal ions) are highly conserved in all legume lectins raises the puzzling question as to how these lectins discriminate between very similar monosaccharides. For example, galactose and glucose differ only in the configuration of the hydroxyl group at C-4 in the carbohydrate ring (axial and equatorial, respectively), yet galactose-specific lectins will not bind glucose (or mannose), nor will those specific for glucose (and mannose) combine with galactose. Still, lectins of either specificity form nearly superimposable hydrogen bonds with this key hydroxyl group through the side-chains of two conserved amino acids, an aspartic acid residue and an asparagine residue,

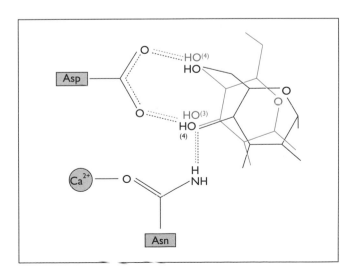

Figure 2. Binding site of legume lectins
Key hydrogen bonds (dashed lines) holding galactose (blue) in the combining site of a galactose-specific lectin (from *E. corralo-dendron*) and glucose (black) in a glucose/mannose-specific lectin (e.g. concanavalin A or pea lectin).

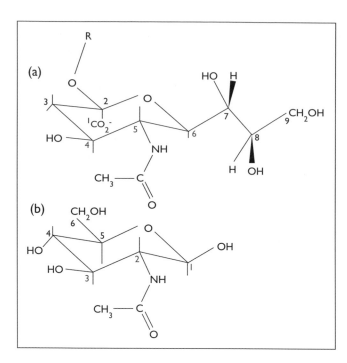

Figure 3. Structure of (a) NeuAc and (b) N-acetylglucosamine
NeuAc is written so that its carbons 4, 5, and 6 are superimposable on carbons 1, 2, and 3 of N-acetylglucosamine.

which also participate in binding the metal ions to the protein (Figure 2). This is possible because the orientation of the monosaccharide in the binding site of the mannose/glucose-specific lectins is different from that in the galactose-specific ones, apparently due to differences in the structure of the variable amino acids that line the binding pocket.

The cereal lectins are very different, as illustrated by wheat germ agglutinin (WGA), the only member of this family to be characterized in molecular detail[14]. WGA is a dimer of two identical 18 kDa subunits and has four carbohydrate-binding sites, located at the interface between the subunits. Each subunit comprises four homologous domains of 43 amino acids; the domains are folded in a similar fashion, each with four identically positioned disulphide bridges. There are thus 16 such bridges per WGA subunit, resulting in a highly stable molecule. The specificity of cereal lectins is somewhat unusual, in that they interact with both sialic acid and N-acetylglucosamine. This can be explained by the similarity in configuration at positions C-2 (acetamido group) and C-3 (hydroxyl group) of the pyranose rings of the two sugars that are critical to productive contact with the WGA combining site (Figure 3). In the crystalline complex of WGA with sialyl-lactose, NeuAc (α2-3)Gal(β1-4)Glc, examined by X-rays at high resolution, the sialic acid is bound to the lectin by a number of hydrogen bonds (Table 1) — none with aspartic acid or asparagine residues, as found in the legume lectins — as well as by non-polar contacts with aromatic amino acids[14]. The amino acids involved in the binding are not located in the same subunit, as found in legume lectins, but belong to two different subunits of the WGA dimer.

Solanaceous lectins (e.g. of tomato and potato), specific for chitin oligo-saccharides (β1-4-linked oligomers of *N*-acetylglucosamine), exist as dimers of two identical subunits. Each subunit consists of two evolutionary autonomous domains: a carbohydrate-binding region fused to a hydroxyproline-rich, highly glycosylated module. The former shares sequence similarities with other chitin-binding plant proteins and, unexpectedly, also with platelet-aggregation inhibitors from snake venoms[15]. The hydroxyproline-rich domain, in turn, is similar to extensins, a family of glycoproteins that are components of plant cell walls. Neither the three-dimensional structure of any of these lectins, nor that of their complexes with sugars, is known.

Animal lectins

The primary structure of approximately 100 animal lectins, mostly from verte-brates, and the three-dimensional structure of two of them, have been determined[16]. While their overall architecture varies widely, carbohydrate-binding activity can often be ascribed to a limited polypeptide segment of each lectin, designated the carbohydrate-recognition domain (CRD). Several types of CRD have been discerned, each of which shares a pattern of invariant and highly conserved amino acid residues at a characteristic spacing. The main ones are the S-type and C-type CRDs; proteins containing such CRDs are known as galectins (previously S-lectins) and C-type lectins, respectively.

Galectins

These lectins bind exclusively β-galactosides, such as lactose [Gal(β1-4)Glc] and *N*-acetyl-lactosamine [Gal(β1-4)GlcNAc], and do not require metal ions for their activity[17]. They occur predominantly in mammals, but have also been described in other species, such as electric eel, nematode and sponges. Rather unusually, they are present both inside the cytoplasm and nucleus of different cells and, occasionally, also on the cell surface and outside the cell. Galectins 1 and 2 are non-covalently bound dimers of identical subunits of approximately 130 amino acids. The subunits are highly homologous, with 19 invariant and 36 conserved residues, that define the S-type CRD. Each subunit folds as one compact, globular domain, similar in its three-dimensional structure to that of the subunits of legume lectins (Figure 4). This is quite remarkable, since there is no significant sequence homology between lectins of these two families, and suggests a common evolutionary origin. Galectins 3 and 4 are monomeric, multidomain molecules with one or two CRDs, respectively. The crystal structures of galectin 1 in complex with lactose, and of galectin 2 in complex with *N*-acetyl-lactosamine, show that of particular importance in the binding of the sugars to the proteins are the hydrogen bonds between the C-4 hydroxyl group of the galactose in these disaccharides and the side-chains of three amino acids (histidine, asparagine and arginine) that are conserved among all galectins (Figure 5a).

Figure 4. Similarity in the overal folding of the lectin from *E. corallodendron* (a) and of bovine spleen galectin I (b), viewed from two different directions
Reproduced with permission from reference 24.

(a) (b)

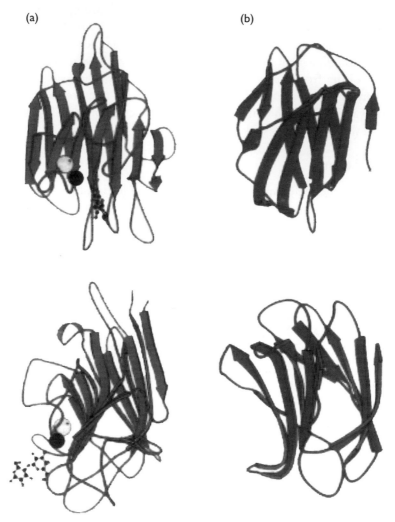

The expression of many galectins is developmentally regulated, i.e. their synthesis in a given tissue is activated only during particular developmental or physiological stages, indicating that they function in important biological processes. However, direct evidence for defined functions is still scanty.

Galectin 1, for example, inhibits interactions between cells and the extracellular matrix in skeletal muscle that may be required in muscle development. Galectin 3 is a component of ribonucleoprotein particles and is concentrated in the nucleus during proliferation of some cell types, consistent with a possible role in RNA processing. Elevated levels of this lectin occur also on the surface of metastatic murine and human cancer cells and it may be responsible for the

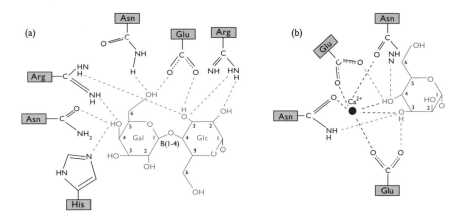

Figure 5. Combining sites of animal lectins
(a) Bovine galectin with bound lactose (blue); (b) soluble mannose-binding protein with bound mannose (blue). Hydrogen bonds and Ca^{2+} coordination bonds are in dashed lines (blue and black, respectively).

adhesion of these cells to their target organs, a step necessary for the formation of metastases. For instance, a good correlation was found between the amount of the lectin expressed on melanoma cells and the formation of pulmonary metastases after injection of the cells into syngeneic mice. Exposing highly metastatic cells to compounds containing lactose before injecting them into the mice reduced the metastatic spread almost by half. These findings indicate that anti-adhesive drugs may turn out to be antimetastatic.

C-Type lectins

This very large group of animal lectins has been so named because the lectins require Ca^{2+} ions for their activity. They occur in serum, extracellular matrix and membranes of vertebrates, and have been found in a few invertebrates as well (e.g. the flesh fly and sea urchin). Their CRDs are different from those of the galectins and consist of a 115–130 amino acid segment, containing 14 invariant and 18 highly conserved amino acid residues. Interestingly, a C-type CRD has also been reported to be present in *B. pertussis* toxin, a bacterial lectin[3]. In contrast with galectins, C-type lectins exhibit a range of specificities: some bind galactose and *N*-acetylgalactosamine; some recognize mannose, *N*-acetylglucosamine and fucose; whereas others combine only with certain oligosaccharides (Table 2). They are also dissimilar in their overall structure, although in all cases their subunits are composite or modular molecules, made up of one or more CRDs, fused to other polypeptide domains; the latter are often homologous to modules found in other extracellular and cell-surface proteins (Figure 1). The C-type lectins are divided into subgroups on the basis of their function and structure.

Endocytic lectins

The first C-type lectins described were the membrane-bound receptors that mediate endocytosis of glycoproteins. Most prominent among them are the mammalian asialoglycoprotein receptor found on hepatocytes, specific for galactose and *N*-acetylgalactosamine; the avian hepatic lectin, specific for *N*-acetylglucosamine; and the fucose- and galactose-specific lectin found on Kupffer cells, the resident macrophages of the liver. They are type II transmembrane proteins, with a short N-terminal cytoplasmic tail, a hydrophobic segment which spans the membrane and an extracellular C-terminal portion, consisting of a neck region and a CDR.

The asialoglycoprotein receptor mediates clearance from the circulation of serum glycoproteins with complex oligosaccharide units from which the terminal sialic acid has been removed, exposing galactose residues. Glycoproteins bound by this receptor are internalized and transported to the lysosomes for degradation. The avian hepatic lectin acts in a similar manner, although binding to this receptor requires that *N*-acetylglucosamine residues of the glycoproteins be exposed.

The mannose-specific macrophage surface lectin differs from the other endocytic C-lectins in that it is a type I transmembrane protein, i.e. its C-terminus is in the cytoplasm and the N-terminal is outside the cell. Moreover, the extracellular region contains eight CRDs, the only known case of a C-type protein with more than one CRD within a single polypeptide. The lectin may participate in antimicrobial defence by mediating lectinophagocytosis (see earlier) of infectious organisms that expose mannose-containing glycans on their surface.

Collectins

Lectins of this group[18,19], represented by the soluble mannose-binding proteins of mammalian serum and liver, are characterized by an N-terminal collagen-like stretch of repeating Gly-Xaa-Yaa triplets (where Xaa and Yaa are any amino acid). The structural unit of the mannose-binding protein is a trimer of identical subunits with a triple-stranded collagen helix and three CRDs (Figure 1). This arrangement of CRDs at a fixed distance has important biological implications, in that it allows the lectin to bind ligands with repetitive, mannose-rich structures such as are found on fungal and microbial surfaces, but not oligomannose chains on mammalian glycoproteins[20].

The three-dimensional structure of the rat mannose-binding protein in complex with an asparaginyl-oligomannose ligand reveals that a Ca^{2+} ion serves as the nucleus of the combining site, forming bonds with hydroxyl groups at C-3 and C-4 of the terminal mannose of the bound oligosaccharide[16]. This unusual role of a metal ion as a direct sugar ligand differs fundamentally from that in the legume lectins, as discussed earlier. Four of the five additional bonds that coordinate the metal to the protein are provided by the side-chains of two glutamic acid and two asparagine residues that also form hydrogen bonds to the same (C-3 and C-4) sugar hydroxyl groups (Figure 5b). The four amino acids just mentioned are conserved in all C-type lectins specific for mannose, two of them in the sequence Glu[185]-Pro[186]-Asn[187]. However, in lectins of this family that recognize galactose instead of mannose, the glutamic acid residue is replaced by glutamine and the asparagine residue by aspartic acid, effectively reversing the position of the side-chain amide and carboxylate groups. On the basis of these observations it was assumed that the side-chain arrangement at the two positions may be a primary determinant of specificity of the C-type lectins. This has indeed been proven by genetic engineering experiments, in which Glu[185] and Asn[187] in the mannose-binding protein were replaced by glutamine and aspartic acid, respectively[20]. The simple switch in position of a single amide group altered the binding specificity of the lectin so that galactose became its preferred ligand.

The collectins function in antibody-independent host defence against a variety of organisms[18,19]. The mannose-binding protein is synthesized in liver and secreted into serum; both processes are regulated as part of the acute phase response — the immediate set of reactions induced by tissue damage (e.g. injury or infection) before the onset of the immune response. The lectin acts by binding to cell-surface oligomannosides of bacteria and fungi, causing lysis of the pathogens. It also enhances phagocytosis of the invading organisms by acting as an opsonin. Clinical evidence for its importance has come from the identification of a mannose-binding protein-deficiency syndrome. It results from a point mutation in one of the collagen-like triplets of the lectins and is associated with recurrent, severe infections in infants, caused by a range of micro-organisms[21].

Selectins

The demonstration that adhesive interactions mediated by surface carbohydrates and selectins, another family of C-type lectins, play a crucial role in leucocyte trafficking to sites of inflammation, and in the migration (homing) of lymphocytes to specific lymphoid organs, was a landmark in lectin research[22]. The selectins (E-, P- and L-selectin) are highly asymmetric, membrane-bound proteins. Their extracellular part consists of an N-terminal CDR, an epidermal growth factor-like domain, and several short repeating units related to complement-binding protein (cf. Figure 1). They bind specifically to the trisaccharide NeuAc(α2-3)Gal(β1-4)[Fuc(α1-3)]GlcNAc, known as sialyl-Lewis[x] (SiaLe[x]; cf. Table 2) and its positional isomer, NeuAc(α2-3)Gal(β1-3) [Fuc(α1-4)]GlcNAc (sialyl-Lewis[a] or SiaLe[a]), with both fucose and sialic acid required for binding; sialic acid can be replaced by another negatively charged group, such as sulphate[23]. The selectins recognize the carbohydrate ligands only when the latter are present on particular glycoproteins, such as cell-surface mucins, pointing to the role of the carrier molecule in lectin–carbohydrate interactions.

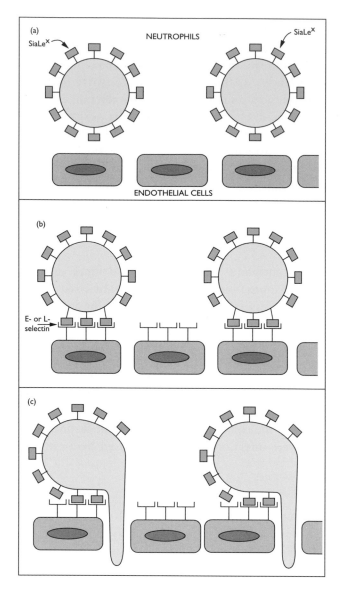

Figure 6.
Oligosaccharide–selectin interactions control leucocyte traffic in the body
(a) Leucocytes, such as neutrophils, move freely in circulation.
(b) Endothelial cells, activated by cytokines, released from infected or otherwise injured tissues, express E- or P-selectins which bind the neutrophils and slow them down.
(c) The slowed-down cells eventually migrate through the blood vessel walls and move to the afflicted site.

The selectins provide the best paradigm for the role of sugar–lectin interactions in biological recognition. In broad outline, they all mediate, although with some differences, the adhesion of circulating leucocytes to endothelial cells of blood vessels, leading to the exit of the leucocytes from the circulation (Figure 6). The extravasation is necessary for the migration of leucocytes into tissues, such as occurs under normal recirculation of lymphocytes between different lymphoid organs, or in recruitment of leucocytes to sites of inflammation. L-selectin, also known as 'homing receptor', is found on all leucocytes. It is involved predominantly in the recirculation of lymphocytes, directing them specifically to peripheral lymph nodes. In contrast with the homing receptor, the two other selectins are expressed mainly on endothelial cells, and only when these cells are activated by inflammatory mediators, mainly cytokines (e.g. interleukin-2 and tumour necrosis factor). The cytokines are released from tissue cells in response to, for example, wounding, infection or ischaemia, and induce the expression of P-selectin on the endothelial surface within minutes, and that of E-selectin within 3–4 h.

The same mechanism which helps leucocytes to breach the endothelial barrier, in the fulfilment of their infection-fighting duties, leads in certain situations to their accumulation in tissues where they should not be, causing tissue damage, swelling and pain: for instance, the inflammation of rheumatoid arthritis. Prevention of adverse inflammatory reactions by inhibition of leucocyte extravasation has become a major aim of many pharmacological industries. In theory, any drug that interferes with the adhesion of white blood cells to the endothelium, and consequently with their exit from the blood vessel, should be anti-inflammatory. However, for anti-adhesive therapy to be successful, the drugs must simultaneously accomplish two seemingly incompatible ends. On the one hand, they must stop leucocytes from leaving the blood stream inappropriately; on the other, they must still allow the cells to go where they are needed. Those goals may be achievable because the specificity of adhesion molecules varies in different tissues. Preliminary experiments in animal models show that oligosaccharides recognized by the selectins do indeed exert protective effects against experimentally induced lung injury.

In addition to their involvement in inflammation, selectins may play a role in the spread of cancer cells from the main tumour throughout the body. This process too requires the migration of cells through the walls of blood vessels. SiaLe^x is expressed on cells from diverse tumours, and at least one type of human cancer cell was demonstrated to bind specifically to E-selectin on activated endothelium. It is possible that, to promote their own metastasis, some malignant cells recruit the adhesion molecules that are part of the body's defence system. If so, inhibitors of selectins may prove to be antimetastatic as well.

Summary

- *Lectins, non-enzymic proteins that bind mono- and oligosaccharides reversibly and with high specificity, occur widely in nature. They come in a variety of sizes and shapes, but can be grouped in families with similar structural features. The combining sites of lectins are also diverse, although they are similar in the same family.*

- *The specificities of lectins are determined by the exact shape of the binding sites and the nature of the amino acid residues to which the carbohydrate is linked. Small changes in the structure of the sites, such as the substitution of only one or two amino acids, may result in marked changes in specificity. The carbohydrate is linked to the protein mainly through hydrogen bonds, with added contributions from van der Waals contacts and hydrophobic interactions. Coordination with metal ions may occasionally play a role too.*

- *Microbial surface lectins serve as a means of adhesion to host cells of viruses (e.g. influenza virus), bacteria (e.g. E. coli) and protozoa (e.g. amoeba): a prerequisite for the initiation of infection. Blocking the adhesion by carbohydrates that mimic those to which the lectins bind prevents infection by these organisms. The way is thus open for the development of anti-adhesive therapy against microbial diseases.*

- *Lectin–carbohydrate mediated interactions between leucocytes and endothelial cells are the first step in the recirculation of lymphocytes and in the migration of neutrophils to sites of inflammation. Such inter-actions may also feature highly in the formation of metastases. Studies of these processes are expected to lead to the development of carbohydrate-based anti-adhesion drugs for the treatment of inflammatory diseases as well as cancer.*

References

References 1, 4, 5, 16, 17 and 22 are particularly recommended for further reading.

1. Sharon, N. & Lis, H. (1989) *Lectins*. Chapman and Hall, London
2. Pusztai, A. (1991) *Plant Lectins*. Cambridge University Press, Cambridge
3. Sandros, J., Rozdzinski, E., Zheng, J., Cowburn, D. & Tuomanen, E. (1994) Lectin domains in the toxin of *Bordetella pertussis*: selectin mimicry linked to microbial pathogenesis. *Glycoconjugate J.* 11, 501–506
4. Sharon, N. (1993) Lectin-carbohydrate complexes of plants and animals: an atomic view. *Trends Biochem. Sci.* 18, 221–226
5. Sharon, N. & Lis, H. (1993) Carbohydrates in cell recognition. *Sci. Am.* 268(1), 82–89
6. Karlsson, K.-A. (1991) Glycobiology: a growing field for drug design. *Trends Pharmacol. Sci.* 12, 265–272
7. Petri, W.A. Jr. (1991) Invasive amebiasis and the galactose-specific lectin of *Entamoeba histolytica*. *ASM News* 57, 299–306
8. Wiley, D.C. & Skehel, J.J. (1987) The structure and function of the hemagglutinating membrane glycoprotein of influenza virus. *Annu. Rev. Biochem.* 56, 365–394
9. Ofek, I. & Sharon, N. (1990) Adhesins as lectins: specificity and role in infection. *Curr. Top. Microbiol. Immunol.* 151, 91–113
10. Morschhauser, J., Hoschutzky, H., Jann, K. & Hacker, J. (1990) Functional analysis of the sialic acid-binding adhesin SfaS of pathogenic *Escherichia coli* by site-specific mutagenesis. *Infect. Immun.* 58, 2133–2138
11. Saukkonen, K., Cabellos, C., Burroughs, M., Prasad, S. & Tuomanen, E. (1991) Integrin-mediated localization of *Bordetella pertussis* within macrophages: role in pulmonary colonization. *J. Exp. Med.* 173, 1143–1149

12. Steuer, M.K.,Beuth, J., Pulverer, G. & Steuer, M. (1994) Experimental and clinical studies on microbial lectin blocking: new therapeutic aspects. In *Lectin Blocking: New Strategies for the Prevention and Therapy of Tumor Metastasis and Infectious Diseases* (Beuth, J. & Pulverer, G., eds.), pp. 112–117, Gustav Fischer Verlag, Stuttgart, Jena, New York

13. Sharon, N. & Lis, H. (1990) Legume lectins — a large family of homologous proteins. *FASEB J.* **4**, 3198–3208

14. Wright, C.S. (1990) 2.2 Å resolution structure analysis of two refined N-acetylneuraminyl-lactose wheat germ agglutinin isolectin complexes. *J. Mol. Biol.* **215**, 635–651

15. Kieliszewski, M.J., Showalter, A.M. & Leykam, J.F. (1994) Potato lectin: a modular protein sharing sequence similarities with the extensin family, the hevein lectin family, and snake venom disintegrins (platelet aggregation inhibitors). *Plant J.* **5**, 849–861

16. Drickamer, K. & Taylor, M.E. (1993) Biology of animal lectins. *Annu. Rev. Cell Biol.* **9**, 237–264

17. Barondes, S.H., Cooper, D.N.W., Gitt, M.A. & Leffler, H. (1994) Structure and function of a large family of animal lectins. *J. Biol. Chem.* **269**, 20807–20810

18. Sastry, K. & Ezekowitz, R.A. (1993) Collectins: pattern recognition molecules involved in first line host defence. *Curr. Opin. Immunol.* **5**, 59–66

19. Holmskov, U., Malhotra, R., Sim, R.B. & Jensenius, J.C. (1994) Collectins: collagenous C-type lectins of the innate immune defense system. *Immunol. Today* **15**, 67–73

20. Weis, W.I. & Drickamer, K. (1994) Trimeric structure of a C-type mannose-binding protein. *Structure* **2**, 1227–1240

21. Sumiya, M., Super, M., Tabona, P. *et al.* (1991) Molecular basis of defect in immunodeficient children. *Lancet* **337**, 1569–1570

22. Lasky, L.A. (1992) Selectins: interpreters of cell-specific carbohydrate information during inflammation. *Science* **258**, 964–969

23. Feizi, T. (1993) Oligosaccharides that mediate mammalian cell-cell adhesion. *Curr. Opin. Struct. Biol.* **3**, 701–710

24. Crennel, S., Garman, E., Laver, G., Vimr, E. & Taylor, G. (1994) Crystal structure of *Vibrio cholerae* neuraminidase reveals dual lectin-like domains in addition to the catalytic domain. *Structure* **2**, 535–544

6

Exocytosis

Alan Morgan

The Physiological Laboratory, University of Liverpool, P.O. Box 147, Liverpool L69 3BX, UK.

Introduction[1,2]

In eukaryotic cells, proteins destined for the plasma membrane or the extracellular space are delivered along the secretory pathway. This comprises a series of sequential, vesicle-mediated transport steps, each of which requires the specific targeting of transport vesicles to the appropriate acceptor membrane and the subsequent fusion of vesicle and acceptor membranes. In this way, proteins to be secreted by the cell are translocated into the endoplasmic reticulum and then travel through the Golgi complex. The proteins are sorted into secretory vesicles in the *trans*-Golgi network and these vesicles then fuse with the plasma membrane. This final membrane fusion event is known as exocytosis and results in the discharge of vesicle contents into the extracellular space as well as the incorporation of vesicle membrane lipids and proteins into the plasma membrane. Exocytosis can be divided into two classes: constitutive and regulated. In constitutive exocytosis, secretory vesicles fuse with the plasma membrane immediately after formation; in regulated exocytosis, secretory vesicles accumulate in the cytoplasm and only undergo fusion upon receipt of an appropriate signal. All eukaryotic cells exhibit constitutive exocytosis, but regulated exocytosis is restricted to certain cells, classic examples being exocrine, endocrine and neuronal cells.

Although the fundamental purpose of exocytosis is to deliver lipids and proteins to the plasma membrane and to release vesicle contents from the cell, different cell types utilize this mechanism to fulfil their own particular physiological role. Some examples of the various functions of exocytosis in different cell types are listed in Table 1 and a schematic diagram illustrating the morphology of some regulated secretory cell types is shown in Figure 1.

Table 1. Functions of exocytosis

Constitutive

All cells	Insertion of plasma membrane proteins
Liver cells	Serum protein secretion
Mammary cells	Milk protein secretion
Fibroblasts	Connective tissue protein secretion

Regulated

Neurons	Neurotransmitter release
Adrenal chromaffin cells	Adrenaline secretion
Pancreatic acinar cells (exocrine)	Digestive enzyme secretion
Pancreatic β-cells (endocrine)	Insulin secretion
Mast cells	Histamine secretion
Mammary cells	Milk protein secretion
Sperm	Enzyme secretion
Egg	Creation of fertilization envelope
Adipocytes	Insertion of glucose transporters into plasma membrane

I have deliberately emphasized regulated secretory cells since most information on exocytosis has come from work on these popular model cells.
Nevertheless, this should not be taken as an indication that exocytosis in these
cells is more important than in others. Indeed, arguably the most important
exocytic fusions are those occurring in sperm, where a single vesicle fuses to
release enzymes which degrade egg vestments and so allow contact with the
egg membrane, and in the unfertilized egg, where cortical granules fuse upon
sperm contact and their released contents form the fertilization envelope
which prevents further sperm access. Thus these two regulated exocytic events
are a prerequisite for the creation of life itself. It should be noted that
exocytosis (i.e. the fusion of vesicles with the plasma membrane) is not only
the end point of the secretory pathway, but can also involve vesicles which did
not originate from the endoplasmic reticulum. For instance, transcytosis
occurs in polarized cells and involves endocytic vesicle budding from one pole
of the cell, transport to the other pole (often via endosomes) and subsequent
exocytic fusion. In mammary cells, transcytosis is used in the uptake of
antibodies from the blood and their subsequent secretion in milk. Similarly,
some vesicles undergo cycles of exo/endocytic fusion via endosomes without
returning to the Golgi. Exocytosis of recycling vesicles may be either
constitutive (e.g. transferrin receptor-containing vesicles) or regulated (e.g.
synaptic vesicles).

Since all cells exhibit constitutive exocytosis, it follows that regulated
secretory cells must posses two types of secretory vesicle: one constitutive and
one regulated. Morphological studies indicate this to be the case, since
constitutive secretory vesicles appear small and clear in the electron
microscope, whereas regulated secretory vesicles typically appear larger and

Figure 1. Schematic diagram of the classic regulated secretory cells
Abbreviations used: N, nucleus; LDCV, large dense-core vesicle; SV, synaptic vesicle.

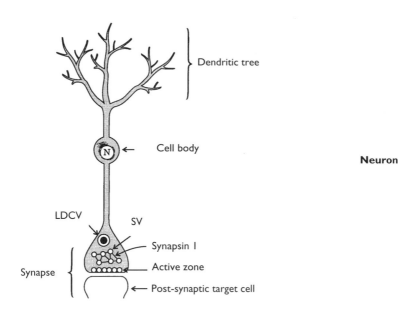

Neuron

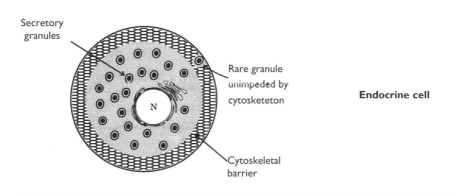

Endocrine cell

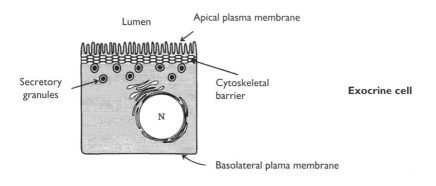

Exocrine cell

Table 2. Characteristics of regulated secretory vesicles in neurons

	Synaptic vesicles	Large dense-core vesicles
Vesicle size	50 nm	200 nm
Speed of transmission	200 μs	milliseconds–seconds
Neurotransmitter	Fast neurotransmitters (GABA, glutamate, ACh, etc.)	Peptides (endorphins, VIP, etc.)
Cells affected	Post-synaptic contact only	Cells in surrounding area
Duration of effect	Short-lived	Longer-lived
Stimulation	Low-frequency	High-frequency
Ca^{2+} concentration	Hundreds of micromolar	Tens of micromolar
Vesicle recycling via endosome?	Yes	No

Abbreviations used: GABA, γ-aminobutyric acid; ACh, acetylcholine; VIP, vasoactive intestinal peptide.

opaque. Furthermore, the two types of vesicle usually contain different substances (an exception is the mammary cell, where casein secretion occurs by both constitutive and regulated exocytosis). It should be noted that cells may contain more than one type of regulated secretory vesicle. The best example of this is seen in neurons, which may possess synaptic vesicles and large dense-core vesicles in addition to constitutive secretory vesicles. Some properties of the two types of neuronal regulated secretory vesicle are listed in Table 2. Large dense-core vesicles contain peptide neurotransmitters and these are very similar to regulated secretory vesicles in endocrine cells. Indeed, much of the information on large dense-core vesicle biogenesis and exocytosis has come from studies of adrenal chromaffin cells and their tumour cell derivatives, PC12 cells, both popular neuronal cell models. Synaptic vesicles appear clear in the electron microscope, are much smaller than large dense-core vesicles and contain fast neurotransmitters. Synaptic vesicles have evolved in animals to allow the extremely rapid point-to-point communication required for brain function. Recently, synaptic-like vesicles have been found in endocrine cells, such as adrenal chromaffin cells and pancreatic β-cells. These vesicles also appear to contain fast neurotransmitters, although their physiological role is unclear.

What triggers regulated exocytosis?

As mentioned above, regulated secretory vesicles fuse with the plasma membrane only upon receipt of an appropriate signal: generally, occupation of plasma membrane receptors or depolarization. A wide variety of receptors trigger exocytosis in different cell types and more than one class of receptor may activate exocytosis in any given cell. For example, although the physiological trigger for adrenaline secretion from adrenal chromaffin cells is acetylcholine, secretion can also be elicited to a lesser extent by bradykinin, angiotensin II and histamine. Despite the bewildering array of extracellular agonists able to elicit regulated secretion, the intracellular signal downstream of receptor occupation which triggers exocytosis is usually an increase in the cytosolic free Ca^{2+} concentration. In some cell types, Ca^{2+} triggers exocytosis by entry from the extracellular space via channels. Neurons are classic examples of such cells, in which depolarization of the axonal plasma membrane by action potentials causes an influx of Ca^{2+} ions into the presynaptic terminal through voltage-dependent Ca^{2+} channels clustered in this region. By contrast, in other cell types, such as pituitary gonadotropes[3], exocytosis is triggered by the release of Ca^{2+} from intracellular stores by inositol 1,4,5-trisphosphate generated in response to receptor activation. The source of the increase in intracellular Ca^{2+} can be crucial in determining whether exocytosis is triggered. For example, in excitable cells, such as adrenal chromaffin cells, release of Ca^{2+} from intracellular stores is unable to evoke adrenaline secretion: Ca^{2+} entry is essential. This is thought to be because the entry of Ca^{2+} through plasma membrane channels allows high concentrations of Ca^{2+} to be achieved locally beneath the plasma membrane, which is the site of exocytic membrane fusion. Indeed, Ca^{2+} imaging studies have revealed that not only the magnitude of the cellular Ca^{2+} rise but also its spatial characteristics determine whether exocytosis is triggered. Thus stimuli which raise Ca^{2+} levels in discrete areas of the cell trigger exocytosis in the same discrete areas. In addition, temporal characteristics of Ca^{2+} signals can be important, since oscillations in intracellular Ca^{2+} concentration have been shown to elicit oscillatory bursts of exocytosis in pituitary cells[3]. Such spatial and temporal control of Ca^{2+} signals is thought to be an adaptation to avoid the cytotoxic effects of prolonged elevations in cellular Ca^{2+} levels.

Although an increase in intracellular Ca^{2+} concentration accompanies exocytosis in most regulated secretory cells, this is not always the case. In parotid acinar cells, for example, salivary enzyme secretion can be elicited by agonists which raise cyclic AMP levels without raising Ca^{2+} concentration, thus implicating protein kinase A as a trigger for exocytosis in these cells. Similarly, phorbol esters (drugs which mimic the action of diacylglycerol by activating protein kinase C) are able to trigger ATP secretion from blood platelets while the resting cellular Ca^{2+} level remains constant, suggesting that protein kinase C can also act as a trigger for exocytosis. It should be remembered that the cytoplasm contains around 100 nM free Ca^{2+} and so the effect of drugs which act in an apparently Ca^{2+}-independent manner may actually be to increase the Ca^{2+} affinity of the exocytic process so that secretion is triggered at resting Ca^{2+} levels. This problem can be overcome using permeabilized cells[4] (cells whose plasma membranes have been selectively porated to allow the exit and entry of normally impermeable macromolecules) incubated with buffers containing high concentrations of calcium chelators to mop up any free Ca^{2+}. Using this approach, the stimulatory effects of cyclic AMP and phorbol esters have been demonstrated in permeabilized parotid acinar cells and pituitary cells at vanishingly low Ca^{2+} concentrations, suggesting a truly Ca^{2+}-independent trigger for exocytosis in some cells.

The pioneering work of Baker and Knight[5] showed that Ca^{2+} could directly trigger exocytosis in permeabilized adrenal chromaffin cells, and similar subsequent studies on other permeabilized cell types have almost without exception confirmed this finding[4]. Typically, maximal secretion was found to occur in response to around 10 μM free Ca^{2+} and required energy in the form of MgATP. However, some permeabilized secretory cells, for example mast cells, are unusual in that a full exocytic response can be elicited in the absence of ATP[4] (see also later). In contrast, while constitutive exocytosis in permeabilized cells also requires MgATP, it is unaffected by Ca^{2+} concentration. Thus Ca^{2+} can be thought of as a ubiquitous intracellular signal for regulated exocytosis (with some exceptions, as noted above). Permeabilized cell studies have also indicated that non-hydrolysable analogues of GTP (such as GTPγS), which act to keep GTP-binding proteins permanently 'switched on', are able to trigger Ca^{2+}-independent exocytosis in many different cell types[4]. This does not mean that GTP is an intracellular signal for exocytosis, however, since GTP levels do not rise upon cell stimulation and since Ca^{2+} triggers exocytosis in the same permeabilized cells. Rather, it indicates that, in addition to a Ca^{2+}-binding protein(s), a GTP-binding protein(s) is also involved in exocytic membrane fusion. The possible identities of these proteins will be discussed later.

Barriers to exocytosis[6,7]

In many regulated secretory cell types, the vast majority of secretory granules are prevented from reaching the plasma membrane by a dense subplasmalemmal cytoskeletal network composed predominantly of actin (Figure 1). In contrast, constitutive secretory vesicles are not impeded by the actin network. The likely explanation for this is that the larger size of regulated secretory vesicles makes them more likely to be obstructed; they also contain several actin-binding proteins which actively tether secretory granules to the cortical cytoskeleton. Clearly, this barrier to exocytosis must be overcome in some way to allow regulated secretory vesicles to fuse with the plasma membrane upon receipt of an appropriate signal. This is achieved by the signal (usually Ca^{2+}) causing the reversible disassembly of the cortical cytoskeleton to allow the vesicles to move to the plasma membrane. Biochemical and microscopical analyses of a wide variety of regulated secretory cells support this theory, and, in some cells, cytoskeletal breakdown occurs in discrete patches that correlate with exocytic release sites. The protein(s) on which Ca^{2+} acts to trigger cytoskeletal disassembly *in vivo* are not known for certain, but there is evidence that scinderin, a Ca^{2+}-dependent, actin-severing protein, is involved in this process in adrenal chromaffin cells. At least one substrate of protein kinase C (possibly scinderin) is likely to be required since phorbol esters are able to cause cytoskeletal disassembly in chromaffin cells. However, as phorbol esters do not elicit secretion in these cells, cytoskeletal disassembly is necessary but not sufficient for regulated exocytosis, and hence Ca^{2+} acts at at least two distinct stages in exocytosis: an early stage of actin rearrangement and a late stage of membrane fusion.

In neuronal exocytosis, a similar mechanism may operate for large dense-core vesicles, but synaptic vesicles are not directly bound to the actin cytoskeleton. Rather, while a small proportion of synaptic vesicles are tightly docked to neurotransmitter release sites at the presynaptic plasma membrane ('the active zone'), the majority of synaptic vesicles are cross-linked by the extrinsic vesicle protein synapsin I, which is itself associated with the actin cytoskeleton (Figure 1). It is not known why synapsin I is specific to neurons, but it may be that the small size of synaptic vesicles means that the cortical actin cytoskeleton does not act as a sufficient barrier, as has been suggested for constitutive secretory vesicles. In response to action potentials, synaptic vesicles at the active zone undergo exocytosis within 200 μs in response to the extremely high (hundreds of micromolar) Ca^{2+} concentrations at the mouth of presynaptic Ca^{2+} channels, to which synaptic vesicles are thought to be attached. Evidently, some mechanism must exist to allow the replenishment of vesicles lost from the active zone as a consequence of such rapid exocytosis. Abundant evidence suggests that this mechanism involves the liberation of vesicles from the reserve pool of synapsin I-linked synaptic vesicles. Ca^{2+} is known to have very limited diffusion in cytosol due to the efficiency of Ca^{2+} binding and extrusion mechanisms; therefore, although the concentration of Ca^{2+} may be several hundred micromolar at the plasma membrane, a short distance within the presynaptic terminal it would be sub-micromolar. Such sub-micromolar Ca^{2+} concentrations activate Ca^{2+}/calmodulin-dependent protein kinase II (EC 2.7.1.123) (present on synaptic vesicles) which phosphorylates synapsin I, leading to its dissociation from the vesicles and so allowing the released synaptic vesicles to move to and dock at the active zone. Synapsin I would then be dephosphorylated by phosphatases to allow cross-linking of new endosome-derived synaptic vesicles. In this way, a steady supply of synaptic vesicles is kept ready for fusion, so that the neuron can respond to repetitive stimulation. Such repetitive (high-frequency) stimulation triggers large dense-core vesicle exocytosis. This is thought to be because these vesicles are located deeper within the nerve terminal and so repeated bursts of Ca^{2+} entry may be required to raise Ca^{2+} to levels sufficient to elicit exocytosis in these areas (Ca^{2+} extrusion lags behind synaptic vesicle exocytosis, leading to 'residual Ca^{2+}'). It should be noted that prolonged high-frequency stimulation depletes the reserve pool of synapsin I-associated vesicles because the rate-limiting step is the recycling of synaptic vesicles via endocytosis and their refilling with neurotransmitter from the cytosol.

Kinetics of regulated exocytosis[8]

As mentioned above, exocytosis of synaptic vesicles can occur within 200 μs of the application of a depolarizing stimulus. Neurotransmission can be studied at such exceptional time resolution by monitoring the electrical response of the post-synaptic target cell. Thus, the entire process of Ca^{2+} entry, triggering of the fusion protein(s), membrane fusion, release of neurotransmitter, diffusion of neurotransmitter to post-synaptic receptors, receptor binding and consequent depolarization of the post-synaptic contact cell occurs within 200 μs. Since non-neuronal cells do not form synapses, it has been difficult to study the kinetics of exocytosis in other regulated secretory cells, until the recent development of patch-clamp capacitance analysis. The patch-clamp technique involves attaching an extremely fine bore micropipette to the surface of a single cell to achieve an electrical seal and then rupturing the portion of membrane under the pipette tip by suction. This results in electrical continuity between the cell cytoplasm and the electrolyte within the micropipette, which allows the electrical properties of the cell to be controlled and measured. Capacitance analysis takes advantage of the directly proportional relationship between plasma membrane surface area and membrane capacitance. Therefore, when a cell is stimulated to secrete, the fusion of secretory vesicle membranes with the plasma membrane results in an increase in plasma membrane area which is detected as a capacitance rise, thus providing an assay of exocytosis in single cells. This method is so sensitive that in cell types with very large secretory vesicles, e.g. *beige* mutant mouse mast cells, capacitance increases corresponding to the fusion of a single granule can be detected.

Patch-clamp studies of regulated secretory cells have revealed that, as in permeabilized cells, exocytosis can be triggered by Ca^{2+} and GTPγS. These studies are performed by placing either Ca^{2+} or GTPγS in the pipette solution so that the trigger gains access to the cytoplasm by diffusion from the pipette tip. This approach is not suitable for accurate kinetic analysis of exocytosis, however, as there is a considerable diffusion lag to be considered. Recently, this problem has been overcome by using flash photolysis of caged Ca^{2+} and GTPγS. Caged compounds are not biologically active but can be activated by photolysis of ultraviolet-labile bonds. Thus, the caged compound is perfused into the cytoplasm through the patch pipette and then uncaged by a very brief flash of ultraviolet light to liberate desired concentrations of the biologically active trigger almost instantaneously. This approach has indicated that the fastest exocytic fusion events occur 2–50 ms after liberation of caged Ca^{2+} in adrenal chromaffin and pituitary cells; and many seconds after liberation of caged GTPγS in mast cells. Thus exocytosis occurs at least 10 times faster in neurons than in endocrine cells and 100000 times faster than in mast cells. The incredible speed of neuronal exocytosis is almost certainly an adaptation to allow the rapid cell–cell communication required for cognitive and motor functions in higher animals. In contrast, inflammatory responses involving mast cells may occur over several minutes and so the need for fast exocytosis is less pressing. Flash photolysis studies have also indicated that the Ca^{2+} affinity of exocytosis is greatly different in neurons and endocrine cells: the rate of exocytosis is half-maximal at 10–20 µM in adrenal chromaffin cells, but 190 µM in retinal neurons. Similarly, intracellular dialysis of Ca^{2+} via the patch pipette indicates that the threshold for activation of exocytosis in chromaffin cells is 0.3 µM, but 20–50 µM in retinal neurons. Thus although chromaffin and pituitary cells are derived from the neural crest during embryogenesis and are extensively used as neuronal cell models, their exocytic responses can be quite different from those of neurons. In particular, the large difference in Ca^{2+} affinity implies that different Ca^{2+}-binding proteins may be involved in exocytosis in neurons and endocrine cells. It may be that exocytosis in endocrine cells accurately reflects large dense-core vesicle exocytosis in neurons but not synaptic vesicle exocytosis.

Flash photolysis studies in pituitary and chromaffin cells have indicated that after the initial burst of fast exocytosis mentioned above, a slower, sustained exocytic response occurs which lasts from seconds to minutes. Experiments on permeabilized chromaffin cells have also shown these two kinetic phases of secretion, albeit with very limited time resolution. The fast phase, lasting from milliseconds to seconds, is a small secretory response representing the fusion of only around 500 vesicles per cell, whereas the slow phase, lasting from seconds to minutes, is a much larger secretory response representing the fusion of around 10000 vesicles per cell. It is thought that these kinetically distinct phases of release are related to the distribution of secretory granules in the cell, since electron microscopical studies have revealed that of the 30000 secretory granules in chromaffin cells, all but around 500 are kept away from the plasma membrane by the cortical actin barrier (see Figure 1). Thus it may be that these 500 vesicles undergo fast exocytosis as a result of their proximity to the plasma membrane, while the requirement for cytoskeletal disassembly retards the exocytic fusion of the remaining vesicles. This may be important *in vivo* since it is likely that only a small number of vesicles are released per chromaffin cell to provoke the 'fight or flight' response (secretion of 10000 vesicles per cell by the whole adrenal medulla would almost certainly release a fatal dose of adrenaline). Therefore, endocrine cells may use a similar mechanism to that employed by neurons to ensure that a limited number of vesicles fuse per stimulus and that a Ca^{2+}-dependent mechanism exists for the replenishment of these vesicles from a reserve pool; however, whereas cross-linking of synaptic vesicles via synapsin I is utilized by neurons, attachment to the cytoskeletal actin barrier is used by endocrine cells.

Work on permeabilized cells has revealed that the two phases of exocytosis in chromaffin cells differ not only in kinetics, but also in requirement for MgATP: fast exocytosis is independent of added MgATP but slow exocytosis requires millimolar levels of MgATP to proceed. These data have been interpreted as providing evidence that exocytosis comprises two sequential biochemical processes: ATP-dependent 'priming' and ATP-independent 'triggering'. Fast exocytosis is said to result from triggering of pre-primed granules and slow exocytosis requires both priming and triggering, thus explaining the different MgATP-dependencies of the kinetic phases of secretion. 'Priming' is a purely functional definition and its molecular mechanism is unknown but could involve any process preceding the Ca^{2+}-triggered membrane-fusion event, including the cytoskeletal rearrangements discussed above. Nevertheless, recent work has suggested that at least part of the priming process is the maintenance of sufficient levels of phosphoinositides in cell membranes (presumably the secretory vesicle and/or plasma membranes). It is not known whether phosphoinositides are required to target phosphoinositide-binding proteins to membranes, or whether they play a secondary messenger-generating role. There are indications that priming is also Ca^{2+} dependent, but with lower levels of Ca^{2+} required (maximal at 1 μM), thus representing yet another site of action of Ca^{2+} in the exocytic process. As mentioned earlier, some cell types (e.g. mast cells) are able to evoke a full exocytic response in the absence of MgATP[4]. It is not known whether in such cells the secretory vesicles exist in a stable, pre-primed state ready for triggering, or whether a fundamentally different exocytic mechanism is in operation.

Proteins involved in exocytosis

Intensive studies of regulated exocytosis over the last 15 years or so have yielded considerable information on the role of Ca^{2+} signalling in exocytosis, but the nature of the proteins which mediate vesicle docking and fusion has remained elusive. Recently, however, a convergence of classical biochemistry and modern molecular genetics has identified several proteins which are involved in exocytosis. Surprisingly, homologues of several of these proteins are also involved in constitutive vesicular fusions throughout the secretory pathway from yeast to man, indicating that the highly specialized machinery for fast exocytosis in synapses has been adapted from an evolutionarily ancient mechanism.

Ca^{2+}-binding proteins

The central role of Ca^{2+} in regulated exocytosis has been recognized for many years, and, therefore, much research has focused on attempts to identify the Ca^{2+}-binding protein(s) which triggers exocytosis. Such studies are complicated by Ca^{2+}-dependent cytoskeletal/synapsin rearrangements and priming processes prior to the Ca^{2+}-triggered fusion step. Although various Ca^{2+}-binding proteins have been nominated, the most likely candidate for the Ca^{2+} receptor in exocytosis is synaptotagmin. Synaptotagmin is a 65 kDa integral membrane protein which is present on synaptic vesicles and endocrine secretory granules. Molecular cloning techniques have revealed that at least four isoforms of synaptotagmin (I–IV) exist, which are differentially expressed in the central nervous system. A role for synaptotagmin in vesicular docking/fusion has been suggested based on its interaction with N-type Ca^{2+} channels, the α-latrotoxin receptor and syntaxin — all presynaptic plasma membrane proteins. These proteins are themselves implicated in neuronal exocytosis: Ca^{2+} channels mediate the Ca^{2+} signal; the α-latrotoxin receptor induces Ca^{2+}-independent exocytosis in response to black widow toxin application; and syntaxin is cleaved by the neurotransmitter-release blocker, botulinum neurotoxin C1.

Synaptotagmin contains a single transmembrane region and two cytoplasmic repeats that are similar to the C_2 domain of protein kinase C, which is thought to govern Ca^{2+}/phospholipid binding. Accordingly, synaptotagmin has been found to bind Ca^{2+} and phospholipid in a ternary complex and the Ca^{2+} affinity of this reaction is such that it is maximal at around 100 μM. The presence of synaptotagmin on the synaptic vesicle suggests that it is the low-affinity Ca^{2+} receptor in fast neurotransmitter release, which is known to require Ca^{2+} concentrations of 100 μM or more[8]. Indeed, a recent report[9] has shown this to be the case. Transgenic mice carrying a loss-of-function mutation in the synaptotagmin I gene died shortly after birth, but cultured neurons from mutant mouse embryos were found to be defective in fast synaptic transmission. However, the slow component of synaptic transmission was unaffected. These data suggest that synaptotagmin I is the low-affinity Ca^{2+} receptor active in fast neurotransmission, but that a different, high-affinity Ca^{2+}-binding protein mediates slower synaptic vesicle exocytosis in response to residual Ca^{2+}.

If synaptotagmin I is the low-affinity Ca^{2+} receptor in exocytosis, what is the high-affinity receptor? This is an important question since, as discussed earlier, dense-core vesicle exocytosis in neurons and secretory granule exocytosis require much lower Ca^{2+} concentrations than synaptic vesicle exocytosis. An obvious possibility is that it is another synaptotagmin isoform. Synaptotagmin III is an attractive candidate for such a protein since PC12 cells expressing this isoform, but lacking synaptotagmins I and II, exhibit normal secretory granule exocytosis. Alternatively, the high-affinity Ca^{2+} receptor may be an entirely different protein. Rabphilin-3A is an extrinsic membrane protein of the synaptic vesicle which is expressed in brain and endocrine tissues. Like synaptotagmin, rabphilin-3A contains two repeats of the C_2 domain of protein kinase C and binds Ca^{2+}/phospholipid; however, rabphilin-3A binds at Ca^{2+} concentrations consistent with a high-affinity Ca^{2+} receptor (maximal at 1–10 μM Ca^{2+})[10]. A role for rabphilin-3A in exocytosis has been suggested based on its binding in a GTP-dependent manner to rab3A, which is itself thought to act in exocytosis (see below), although it must be stressed that no functional evidence yet exists for an action of rabphilin-3A in exocytosis.

The absence of synaptotagmin and rabphilin expression in exocrine cells means that these proteins cannot be ubiquitous Ca^{2+} receptors in all secretory cells, but several other Ca^{2+}-binding proteins could potentially fulfil this role. p145 is a soluble protein which was purified based on its ability to stimulate micromolar Ca^{2+}-triggered secretion from permeabilized PC12 cells[1]. It appears to be a Ca^{2+}/phospholipid-binding protein, since its purification involves Ca^{2+}-dependent binding to hydrophobic surfaces, although its Ca^{2+} affinity is unknown. Intriguingly, p145 is expressed in most regulated secretory cells (including brain, endocrine and exocrine tissues) but not in constitutive secretory cells[1]. The annexins are a family of Ca^{2+}/phospholipid-binding proteins which are widely expressed. Annexins aggregate purified chromaffin granules in the presence of Ca^{2+}, and this effect of annexin II is half-maximal at 1.8 μM Ca^{2+} (interestingly, half-maximal stimulation of exocytosis in permeabilized chromaffin cells occurs at 1 μM Ca^{2+}). Functional evidence for a role of annexin II in exocytosis comes from the observation that the purified protein stimulates Ca^{2+}-triggered exocytosis in permeabilized adrenal chromaffin cells[1]. Annexin II has also been shown to be required for Ca^{2+}-dependent endocytic vesicle fusion, and so annexins may be involved in membrane fusion events at multiple stages of the secretory pathway. Finally, functional evidence supports a role for calmodulin in exocytosis, since the purified protein stimulates secretion in permeabilized chromaffin cells and since vesicle docking is blocked in calmodulin mutants of the protozoa *Paramecium*[11]. Calmodulin is widely expressed and its Ca^{2+}-dependency (half-maximal at 1–10 μM Ca^{2+}) is consistent with the predicted properties of the high-affinity Ca^{2+} receptor in exocytosis. However, the lack of effect of calmodulin-blocking drugs on secretion in some permeabilized cells means that the role of calmodulin in regulated exocytosis remains controversial.

GTP-binding proteins

Permeabilized cell and patch-clamp capacitance studies have shown that non-hydrolysable GTP analogues, such as GTPγS, trigger exocytosis in many cell types, thus implicating GTP-binding proteins in exocytosis[4]. GTP-binding proteins can be divided into two classes: monomeric GTP-binding proteins and heterotrimeric G-proteins, and there is evidence that both classes of protein are involved in exocytosis and indeed in vesicular transport in general. The first GTP-binding protein to be identified in the control of vesicular transport was the monomeric GTP-binding protein, Sec4p. Yeast with a loss-of-function mutation in the *SEC4* gene accumulate secretory vesicles due to a block of constitutive exocytosis. Cloning studies have revealed that Sec4p has sequence similarity with a family of over 20 mammalian monomeric GTP-binding proteins, which are collectively known as rab proteins. Each rab protein has a distinctive subcellular localization, leading to speculation that rabs act to ensure the correct targeting of vesicles to acceptor membranes in the secretory and endocytic pathways. This theory is supported by observations that rab mutants block membrane traffic at the point in the pathway to which the specific rab is localized. For example, rab5 is localized to endosomal and plasma membranes and rab5 mutants block endocytosis. Regarding exocytosis, rab11 is present on both constitutive and regulated secretory vesicles, while rab3 (which has four isoforms, A–D) is localized to regulated secretory vesicles only. Functional studies have shown that attenuation of rab3B expression using antisense oligonucleotides — which anneal to rab3B mRNA, leading to its destruction by RNase H — blocks exocytosis in pituitary cells, suggesting an essential role for this protein in regulated secretion. Further evidence for a role of rab3 in exocytosis comes from the observation that chromaffin cells transfected to overexpress wild-type or mutant rab3A showed reduced secretion in both intact and permeabilized cells. In contrast, transgenic mice carrying a loss-of-function mutation in the rab3A gene had little effect on synaptic transmission. However, such 'knockout mouse' experiments, though elegant, should be treated with caution since there may be functional overlap between different proteins, and so it may be that, in the absence of rab3A, another rab3 isoform can substitute. Taken together, the evidence favours a role of rab3 in regulated exocytosis[12], although whether its role is in targeting or as a checkpoint control preceding membrane fusion remains to be demonstrated.

Heterotrimeric G-proteins have been widely studied in intracellular signalling as they couple many receptors to their intracellular effector proteins. Indeed, G-proteins have an essential signal-transducing role in many regulated secretory cells. Around 10 years ago, it was suggested that a G-protein (termed G_E, for exocytosis) was a component of the fusion apparatus, since activation of exocytosis in permeabilized mast cells by GTPγS was independent of known secondary messenger generation[4]. G_E appeared to be a G-protein since AlF_4^- mimicked the effect of GTPγS in stimulating secretion; however, AlF_4^- is unable to activate small GTP-binding proteins. Following this work, it has been suggested that G-proteins may play a role in a number of vesicular transport steps, including exit from the Golgi stack and endocytic vesicle fusion. Recently, it has been suggested that the identity of G_E in mast cells is G_{i3}, based on the observation that synthetic peptides corresponding to the C-terminus of G_{i3} (and antibodies raised against them) inhibit exocytosis in permeabilized mast cells. Similar peptide and antibody experiments have indicated that G_o exerts a tonic inhibitory effect on secretion from permeabilized chromaffin cells[1]. Thus both stimulatory and inhibitory G-proteins may be involved in regulated exocytosis. Although it has been suggested that they may control exocytosis by interacting with the membrane fusion apparatus directly, it remains possible that they activate an as yet unknown intracellular signalling pathway.

SNAREs, SNAPs and NSF

Tetanus and botulism are fatal diseases caused by extremely potent neurotoxin proteins released by *Clostridium tetani* and *C. botulinum*, respectively. Tetanus toxin and the seven botulinum neurotoxins (designated A–G) were known to poison nerve terminals by blocking neurotransmitter release at an intracellular site. However, their mechanism of action was mysterious until recently, when the pioneering work of Schiavo *et al.*[13] revealed that tetanus and botulinum B toxins are zinc-dependent endoproteases that block neuro-transmitter release by the specific cleavage of the synaptic vesicle protein, synaptobrevin (also known as vesicle-associated membrane protein; VAMP). This discovery prompted a rush to discover the proteolytic substrates for the other botulinum toxins, which were subsequently identified as synaptobrevin, SNAP-25 (synaptosomal-associated protein, 25 kDa) and syntaxin (also known as HPC-1) (see Table 3; reviewed in reference 14). Thus, the synaptic vesicle protein synaptobrevin and the presynaptic plasma membrane proteins SNAP-25 and syntaxin are essential for synaptic vesicle exocytosis. The clostridial neurotoxins also inhibit secretion in some endocrine cells (e.g. adrenal chromaffin cells), indicating that these three proteins are also essential for exocytosis in non-neuronal cells. Indeed, isoforms of synaptobrevin have recently been identified on various regulated secretory vesicles, including zymogen granules in pancreatic acinar cells and glucose transporter-containing vesicles in adipocytes, and also on constitutive recycling vesicles. The importance of synaptobrevin-like proteins in vesicular fusion is underscored by the observation that its homologues (Snc1p and Snc2p) are present on post-Golgi vesicles and are required for exocytosis of these vesicles in yeast. Yeast homologues of syntaxin (Sso1p and Sso2p) and SNAP-25 (Sec9p) are also thought to function in exocytosis, indicating that similar vesicle and plasma membrane proteins mediate docking and/or fusion of secretory vesicles in yeast and mammalian brain. Furthermore, other yeast homologues of synapto-brevin (Bos1p, Bet1p and Sec22p) and syntaxin (Sed5p) are required for fusion of endoplasmic reticulum-derived vesicles with the *cis*-Golgi, implying that the same protein families are involved in all vesicular fusion processes (reviewed in reference 15).

Table 3. Clostridial neurotoxin substrates

Neurotoxin	Target
Tetanus	Synaptobrevin
Botulinum B	
Botulinum D	
Botulinum F	
Botulinum G	
Botulinum A	SNAP-25
Botulinum E	
Botulinum C1	Syntaxin

Table 4. Proteins for which functional evidence supports a role in exocytosis

Protein	Role	Functional evidence
Annexin II[1]	?	Permeabilized cells
Calmodulin[11]	Vesicle docking in *Paramecium* ?	Molecular genetics Permeabilized cells
Cysteine string proteins[19]	Ca^{2+} channel subunits	Molecular genetics
G-proteins[4]	?	Permeabilized cells
GAP-43[1]	? ?	Permeabilized synaptosomes Molecular genetics
Neurexins[1]	?	Toxicology (α-latrotoxin)
p145[1]	?	Permeabilized cells
Phosphatidylinositol transfer protein[23]	'Priming'	Permeabilized cells
Protein kinases[1]	Cytoskeletal changes, ?	Pharmacology
Protein phosphatases[1]	?	Pharmacology
Rab3[12]	? ?	Patch-clamp capacitance Permeabilized cells
Rop[22]	Vesicle docking?	Molecular genetics
SNAP-25[14]	Vesicle docking/fusion?	Toxicology (botulinum toxins) Molecular genetics
Synaptobrevin[14]	Vesicle docking/fusion?	Toxicology (botulinum toxins)
Synaptotagmin[9]	Vesicle Ca^{2+} receptor	Molecular genetics
Syntaxin[14]	Vesicle docking/fusion?	Toxicology (botulinum toxins) Molecular genetics
SNAPs[18]	Vesicle docking/fusion?	Permeabilized cells
14-3-3 proteins[1]	?	Permeabilized cells

In apparently unconnected work, a search for cytosolic proteins involved in vesicular fusion in the Golgi stack *in vitro* identified NSF (*N*-ethyl-maleimide-sensitive fusion protein) and SNAPs (soluble NSF-attachment proteins) as essential proteins (reviewed in reference 16). These proteins are thought to function in membrane fusion events at all stages of the constitutive secretory and endocytic pathways, and homologues of NSF (Sec18p) and α-SNAP (Sec17p) are required for vesicular transport in yeast[16]. NSF and SNAPs were known to assemble into a 20S complex in the presence of detergent-solubilized membranes, prompting a search for the membrane receptors for SNAPs. Recently, in a paper by Sollner *et al.*[17] that shook the field, the SNAP receptors (SNAREs) in brain membranes were identified as synaptobrevin, syntaxin and SNAP-25. Since these are all substrates for the clostridial neurotoxins, this suggested that NSF and SNAPs also acted in the tightly regulated process of neurotransmitter release in addition to their known function in constitutive membrane fusion. Indeed, SNAPs have recently been shown to stimulate Ca^{2+}-dependent exocytosis in permeabilized adrenal chromaffin cells[18]. Synaptobrevin, syntaxin and SNAP-25 can be isolated in a 7S complex to which α-SNAP, and subsequently NSF, can bind, suggesting a sequential process in which the SNAREs act at an earlier stage in exocytosis than SNAPs and NSF[19]. It has been proposed[16,19] that the inter-action of the synaptic vesicle protein synaptobrevin with the plasma membrane proteins syntaxin and SNAP-25 forms a docking complex which acts to target synaptic vesicles to neurotransmitter-release sites at the active zone, and that the general fusion proteins, SNAPs and NSF, subsequently bind to the complex to effect ATP-hydrolysis-driven membrane fusion as a consequence of the intrinsic ATPase activity of NSF. This explicit model, known as the SNARE hypothesis, can be applied to any vesicular transport step and so has been enthusiastically embraced by cell biologists, to the degree where it has become an accepted dogma. However, various observations cast doubt on the model: for example, exocytosis occurs normally in mast cells in the absence of ATP hydrolysis[4], and recent evidence favours a role for synap-tobrevin in vesicle fusion but not docking[20]. Nevertheless, while the precise details of the model may be inaccurate, there is little doubt that SNAREs, SNAPs and NSF are of central importance in vesicular fusion throughout the secretory and endocytic pathways.

Future goals: desperately seeking fusion

The recent explosion of information on the identities of proteins involved in vesicular transport has profoundly altered popular conceptions of the molecular mechanism of exocytosis. Whereas only a few years ago it was thought by many that a single Ca^{2+}-activated, membrane-binding protein might control regulated exocytosis, it is now clear that the co-ordinated action of a large number of proteins is required for the precise targeting of vesicles to membranes and their subsequent fusion. Indeed, for some of the proteins discussed above, theories have been put forward to explain how individual proteins can interact with one another to form a multisubunit docking/fusion machine. The picture is far from complete, however, since there is functional evidence for the involvement of several other proteins in exocytosis (see Table 4). Mechanisms of action can be confidently predicted for some of these proteins. For example, cysteine string proteins are synaptic vesicle proteins which act as subunits/modulators of presynaptic Ca^{2+} channels, and are essential for neurotransmitter release *in vivo*[21]. It is likely that cysteine string proteins perform a dual function in docking synaptic vesicles at Ca^{2+} channels in the active zone and simultaneously enabling the channels to mediate Ca^{2+} flux in response to depolarization. For most other proteins, good evidence for a mechanism of action is lacking. For example, Rop (also known as Munc-18, mSec1, rbSec1 and nSec1), which is homologous to yeast Sec1p, has been suggested to be involved in vesicle docking due to its interaction with syntaxin, but its precise function is unknown[22]. A major challenge for the future is to understand at the molecular level how this latter group of proteins (and, no doubt, others which are as yet unidentified) control exocytosis.

Despite the identification of proteins involved in exocytosis, one fundamental question remains unanswered, namely, which protein catalyses membrane fusion? Although fusogenic properties have been inferred for some proteins (e.g. synaptobrevin) there are no convincing data. Ultimately, the answer to this question may not be solved until membrane fusion can be reconstituted in the test tube using artificial liposomes and recombinant proteins. Even then, it may be that no single protein acts as a fusogen, but rather that a scaffold of many proteins acts to pull the lipid bilayers of each membrane together within the confines of the scaffold. Indeed, this latter theory is supported by patch-clamp capacitance analysis of exocytosis of large secretory vesicles, where lipid flows through the fusion pore (the initial 'hole' connecting the vesicle lumen with the extracellular space) during transient fusion events. Regardless of whether fusion is catalysed by a single fusogen or a scaffold, the lipid composition of the vesicle and target membranes is likely to be critical, and lipid modifications may well be required to allow fusion. Unfortunately, the role of lipids in vesicular fusion remains the orphan of research into the secretory pathway and will probably only become apparent after all of the essential proteins have been identified. Nevertheless, given the rate of progress in the study of vesicular transport, it seems likely that by the year 2000 we will have a clear understanding of the molecular mechanism of exocytic membrane fusion.

Summary

- *Exocytosis is the fusion of secretory vesicles with the plasma membrane and results in the discharge of vesicle content into the extracellular space and the incorporation of new proteins and lipids into the plasma membrane.*

- *Exocytosis can be constitutive (all cells) or regulated (specialized cells such as neurons, endocrine and exocrine cells). Regulated exocytosis is usually, but not always, triggered by an increase in the cytosolic free Ca^{2+} concentration.*

- *In neurons and endocrine cells, a small proportion of regulated secretory vesicles are ready to fuse with the plasma membrane in response to cell stimulation, but the majority are kept in reserve for subsequent stimulation by linkage to a filamentous network of synapsins (in neurons) or actin (in endocrine cells).*

- *Regulated exocytosis varies greatly in kinetics and Ca^{2+} dependency between cell types.*

- *It is likely that several different Ca^{2+}-binding proteins are involved in regulated exocytosis, with synaptotagmin apparently essential for fast exocytosis at synapses.*

- *GTP-binding proteins of both the monomeric and heterotrimeric forms are involved in exocytosis, although their precise role is unclear.*

- *Intense current interest focuses on the idea that the molecular mechanism of vesicle docking and fusion is conserved from yeast to mammalian brain. The SNARE hypothesis postulates that vesicle SNAREs (synaptobrevin and homologues) mediate docking by binding to target SNAREs (syntaxin/SNAP-25 and homologues), whereupon SNAPs and NSF bind to elicit membrane fusion.*

References

1. Burgoyne, R.D. & Morgan, A. (1993) Regulated exocytosis. *Biochem. J.* **293**, 305–316
2. Bennett, M.K. & Scheller, R.H. (1994) A molecular description of synaptic vesicle membrane trafficking. *Annu. Rev. Biochem.* **63**, 63–100
3. Tse, A., Tse, F.W., Almers, W. & Hille, B. (1993) Rhythmic exocytosis stimulated by GnRH-induced calcium oscillations in rat gonadotropes. *Science* **260**, 82–84
4. Lindau, M. & Gomperts, B.D. (1991) Techniques and concepts in exocytosis: focus on mast cells. *Biochim. Biophys. Acta* **1071**, 429–471
5. Baker, P.F. & Knight, D.E. (1978) Calcium-dependent exocytosis in bovine adrenal medullary cells with leaky plasma membranes. *Nature (London)* **276**, 620–622
6. Trifaró, J.-M. & Vitale, M.L. (1993) Cytoskeleton dynamics during neurotransmitter release. *Trends Neurosci.* **16**, 466–472
7. Valtorta, F., Benfenati, F. & Greengard, P. (1992) Structure and function of the synapsins. *J. Biol. Chem.* **267**, 7195–7198
8. Burgoyne, R.D. & Morgan, A. (1995) Calcium and secretory vesicle dynamics. *Trends Neurosci.*, **18**, 191–195

9. Geppert, M., Goda, Y., Hammer, R.E., *et al.* (1994) Synaptotagmin I: a major Ca^{2+} sensor for transmitter release at a central synapse. *Cell* **79**, 717–727

10. Yamaguchi, T., Shirataki, H., Kishida, S., *et al.* (1993) Two functionally different domains of rabphilin-3A, Rab3A p25/smgp25A-binding and phospholipid- and Ca^{2+}-binding domains. *J. Biol. Chem.* **268**, 27164–27170

11. Kerboeuf, D., Le Berre, A., Dedieu, J.-C. & Cohen, J. (1993) Calmodulin is essential for assembling links necessary for exocytotic membrane fusion in *Paramecium*. *EMBO J.* **12**, 3385–3390

12. Lledo, P.M., Johannes, L., Vernier, P., *et al.* (1994) Rab3 proteins: key players in the control of exocytosis. *Trends Neurosci.* **17**, 426–432

13. Schiavo, G., Benfenati, F., Poulain, B., *et al.* (1992) Tetanus and botulinum B neurotoxins block neurotransmitter release by proteolytic cleavage of synaptobrevin. *Nature (London)* **359**, 832–835

14. Montecucco, C. & Schiavo, G. (1994) Mechanism of action of tetanus and botulinum neurotoxins. *Mol. Microbiol.* **13**, 1–8

15. Ferro-Novick, S. & Jahn, R. (1994) Vesicle fusion from yeast to man. *Nature (London)* **370**, 191–193

16. Rothman, J.E. (1994) Mechanisms of intracellular protein transport. *Nature (London)* **372**, 55–63

17. Söllner, T., Whiteheart, S.W., Brunner, M., *et al.* (1993) SNAP receptors implicated in vesicle targeting and fusion. *Nature (London)* **362**, 318–324

18. Morgan, A. & Burgoyne, R.D. (1995) A role for soluble NSF attachment proteins (SNAPs) in regulated exocytosis in adrenal chromaffin cells. *EMBO J.* **14**, 232–239

19. Söllner, T., Bennett, M.K., Whiteheart, S.W., Scheller, R.H. & Rothman, J.E. (1993) A protein assembly–disassembly pathway *in vitro* that may correspond to sequential steps of synaptic vesicle docking, activation and fusion. *Cell* **75**, 409–418

20. Hunt, J.M., Bommert, K., Charlton, M.P., *et al.* (1994) A post-docking role for synaptobrevin in synaptic vesicle fusion. *Neuron* **12**, 1269–1279

21. Umbach, J.A., Zinsmaier, K.E., Eberle, K.K., Buchner, E., Benzer, S. & Gunderson, C.B. (1994) Presynaptic dysfunction in Drosophila *csp* mutants. *Neuron* **13**, 899–907

22. Harrison, S.D., Broadie, K., van de Goor, J. & Rubin, G.M. (1994) Mutations in the Drosophila *rop* gene suggest a function in general secretion and synaptic transmission. *Neuron* **13**, 555–566

23. Hay, J.C. & Martin, T.F.J. (1993) Phosphatidylinositol transfer protein required for ATP-dependent priming of Ca^{2+}-activated secretion. *Nature (London)* **366**, 572–575

7

Intracellular calcium channels

Richard H. Ashley

Department of Biochemistry, University of Edinburgh, George Square, Edinburgh EH8 9XD, U.K.

Introduction

Calcium is present in the cytoplasm of all cells. Its ionized concentration ranges over at least two orders of magnitude from about 100 nM (representing 'resting' or unstimulated levels) up to about 10 μM (only seen transiently, following 'stimulation'). Free cytoplasmic Ca^{2+} modulates the functions of key cellular proteins (e.g. by promoting phosphorylation or dephosphorylation, or by directly modifying protein–protein interactions). Dramatic actions include the initiation of neurotransmitter and neurohormone release (excitation–secretion coupling), and excitation–contraction (EC)-coupling. Cells therefore regulate calcium concentration ($[Ca^{2+}]$) very closely. They normally maintain their resting cytoplasmic $[Ca^{2+}]$ at a level which is, in effect, some 10000-fold below extracellular $[Ca^{2+}]$ by having a very low plasma membrane permeability for the cation, by active Ca^{2+}-extrusion, and by buffering cellular Ca^{2+} using (relatively) mobile binding proteins and (relatively) immobile membrane-bound stores (e.g. the endoplasmic reticulum, ER).

Intracellular Ca^{2+} stores have themselves come to assume important functions in cells. For example, while the Ca^{2+} required for fast neurotransmitter release at the plasma membrane enters from outside, the Ca^{2+} involved in EC-coupling in striated and cardiac muscle cells must reach the contractile apparatus deep within the myocyte. In this case it is largely provided from the sarcoplasmic reticulum (SR). In fact, Ca^{2+} is often released from a nearby store in critical regions of many cells, greatly reducing the effects of intracellular buffering and avoiding the extra metabolic cost which would be involved in the subsequent export of extra Ca^{2+} acquired from outside. Intracellular Ca^{2+}

storage has recently been reviewed[1], and this essay will concentrate on the intracellular ion channels which regulate Ca^{2+} release.

New approaches to study intracellular Ca^{2+}-handling

Advances in a particular field are often associated with the introduction of powerful new techniques. Our current understanding of the molecular basis of the regulated release of Ca^{2+} from intracellular stores would have been impossible without the development of methods for the reconstitution of largely inaccessible intracellular ion channels in voltage-clamped planar lipid bilayers[2]. Similarly, our current appreciation of the dynamic role of intracellular Ca^{2+} in so many cellular processes owes much to the recent invention (and subsequent elaboration) of novel chemically synthesized fluorescent Ca^{2+}-indicators[3].

Fluorescent Ca^{2+} probes

These are now very widely used to obtain critical data (see reference 1), although their sequestration into intracellular organelles still occasionally leads to inappropriate estimates of 'cytoplasmic' $[Ca^{2+}]$. Also, when considering how probe or dye responses occur and how the probes are calibrated, it is clear that under some circumstances the dyes themselves may significantly buffer cytoplasmic Ca^{2+}-transients. Figure 1 illustrates responses from Fluo-3 which, at room temperature, has a dissociation constant for Ca^{2+} (k_d) of ~400 nM. Other probes include 'ratio' dyes such as for Indo-1: in this case, the fluores-

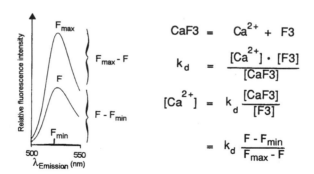

Figure 1. Calibration of Fluo-3 fluorescence

Fluorescence emission scans after exciting Fluo-3 (F3) at 490 nm, showing signals under resting (F), Ca^{2+}-saturated (F_{max}) and Ca^{2+}-free (F_{min}) conditions. Assuming 1:1 stoicheiometry, free $[Ca^{2+}]$ is given by the law of mass action as shown. k_d, the dissociation constant of the Ca^{2+}–dye complex (CaF3), may be determined independently in titration experiments (it is around 400 nM at room temperature, which is the free $[Ca^{2+}]$ shown). The total [dye] = [CaF3]+[F3], while the ratio [CaF3]/[F3] is clearly $(F-F_{min})/(F_{max}-F)$. As 'resting' cytosolic $[Ca^{2+}]$ is ~100 nM, entrapment of, say, 50 μM Fluo-3 will buffer approximately 10 μM additional Ca^{2+} (which must be stripped from internal stores and/or enter from outside the cell). For 'ratio' dyes, the emission (or excitation) spectrum shows a clear shift on binding Ca^{2+}, and the isobestic point can be used as a fixed point of reference.

cence emission spectrum shifts on Ca^{2+}-binding, but the basic principles of dye calibration are the same. Such dyes are now commonly used with a confocal fluorescence microscope and a digital imaging system to compute false-colour representations of regional [Ca^{2+}] in single cells. In this way it is possible to follow local changes in [Ca^{2+}] evolving over tens or hundreds of milliseconds.

(a)

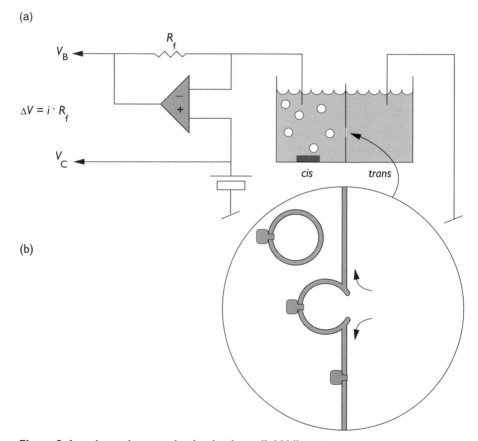

(b)

Figure 2. Ion channel reconstitution in planar lipid bilayers
(a) A planar lipid bilayer, containing membrane phospholipids with or without dispersants such as n-decane is formed across a small (0.2–0.5 mm diameter) hole in a plastic partition separating two solution-filled chambers. In this configuration, the *cis* chamber is voltage-clamped at a given potential (V_C) relative to the *trans* chamber, which is grounded, by using an operational amplifier (opamp) configured as a current-to-voltage converter. When small currents pass through the bilayer (e.g. through incorporated ion channels) the opamp attempts to keep the potentials at its non-inverting (+) and inverting (−) inputs the same, i.e. set to V_C, by adjusting the output or backoff potential (V_B) applied to the feedback resistor R_f (which is commonly 10 GΩ). According to Ohm's law, the (virtual) transmembrane currents are then given by (minus) $[V_B - V_C]/R_f$. A stirbar and 'vesicles' are also shown in the *cis* chamber. (b) Membrane/vesicle fusion is illustrated diagrammatically to emphasise how the orientation of ion channels is maintained relative to their disposition in the original vesicle. Note that the bilayer is 100–1000-fold wider than most membrane vesicles.

Single-channel reconstitution

Figure 2 shows diagrammatically how purified membrane vesicles — prepared, for example, from a subfraction of muscle SR — can be induced to fuse with a preformed planar lipid bilayer in the presence of a *cis>trans* osmotic gradient (although complete understanding of the molecular mechanism of vesicle–bilayer fusion remains elusive). Note that the orientation of membrane proteins is conserved, and in fact the SR membrane vesicles which fuse with the bilayer are 'right side out', so that the cytoplasmic (myoplasmic) sides of incorporated ion channels face the *cis* chamber. If the bilayer is voltage-clamped, as shown, small currents passing through the very-high-resistance (about 100 GΩ) membrane in an all-or-none fashion as an ion channel opens and closes are compensated for electronically by an adjustment in the backoff potential (according to Ohm's law, voltage = current × resistance: 10 mV per pA in Figure 2). These fluctuations can be observed and recorded, after low-pass filtering to reduce high-frequency electrical noise. The 'unit currents' are typically about 1–10 pA ($1 \, pA = 10^{-12} \, A$) in amplitude, depending on the specific channel and the membrane potential and ion concentration difference (i.e. the electrical and chemical driving forces). Note that the constant 'background' or leakage current through the bilayer itself is electronically subtracted. The major significance of 'single-channel recording', especially attractive in a biochemical context, is that it provides a unique opportunity to monitor the functional activity of a single protein molecule.

The functional reconstitution of single ion channels, and fluorescence measurements of the dynamics of regional $[Ca^{2+}]$ in single cells, have recently been combined with a variety of biochemical, molecular biological and structural techniques to greatly extend our understanding of the molecular mechanisms of Ca^{2+} release. It is now clear that two main classes of distantly related ion channel, one sensitive to ryanodine and the other to inositol 1,4,5-trisphosphate $[Ins(1,4,5)P_3]$, are central to this process. In addition, many other channels may directly or indirectly influence intracellular Ca^{2+} release.

Ryanodine-sensitive Ca^{2+}-release channels

Functional reconstitution and purification

Planar bilayers were employed in a crucial series of experiments to reconstitute Ca^{2+} channels from rabbit skeletal muscle SR membrane vesicles[4]. Similar channels were soon discovered in cardiac muscle SR. Channel behaviour was subsequently shown to be modified by the toxic plant alkaloid ryanodine, a commercially exploited insecticide known to disrupt EC-coupling by interfering with intracellular Ca^{2+} release. Ryanodine, at cytoplasmic concentrations of 1–10 μM, appears to alter both the 'gating' (opening and shutting) of the channel, and also its conductance. The mechanism by which this occurs is not yet understood in any great detail, but it does provide a distinctive channel 'fingerprint'. The discovery of ryanodine-sensitive Ca^{2+}-release channels in both vascular[5] and visceral[6] smooth muscle SR might have been anticipated, given the suspected role of intracellular ryanodine-sensitive Ca^{2+}-release channels in EC-coupling, but it was surprising to find the channels were also present in brain[7], and many other non-contractile tissues. Their ubiquity immediately suggested an important general role in Ca^{2+} signalling, similar to the $Ins(1,4,5)P_3$-activated channels described later.

Functional channel proteins were purified using tritiated ryanodine as a specific high-affinity (k_d ~1–10 nM) marker. The solubilized protein complex (ryanodine receptor, RyR) had an unusually high molecular mass (estimated from its sedimentation coefficient to be ~2000 kDa), and astute use of this physical property[8] permitted extensive one-step purification by size-fractionation on a sucrose density gradient. The presence of a single polypeptide band of ~500 kDa on reducing SDS/PAGE suggested that the channel protein was assembled from four large homo-oligomers. This conclusion was supported by measurements of ryanodine binding capacity, provided it was assumed that there was only one high-affinity binding site per tetramer. The channel did not appear to be glycosylated, and the functional properties of the purified protein were broadly similar to those of the native receptor[8,9]. However, it has recently become apparent that channel behaviour is modulated by non-covalently associated accessory proteins. These may not always be removed during the isolation of native (or recombinant) channels. 'Single channel' properties may, therefore, have to be interpreted with more caution, particularly as the accessory proteins appear to have significant roles related to channel phosphorylation and in setting the 'sensitivity' of the channel to physiological ligands *in vivo* (see later).

Reconstituted RyRs often display voltage-sensitive gating, especially the 'type-1' channels isolated from skeletal muscle SR. This raises the possibility that the channels may be influenced by the plasma (sarcolemmal) membrane potential. In addition, the activity of many of the channels, especially the 'type-2' channels isolated from cardiac SR, is modulated by myoplasmic Ca^{2+}, suggesting a molecular mechanism for the phenomenon of the 'Ca^{2+}-induced Ca^{2+} release' that is seen especially clearly in heart. Unlike Ca^{2+}, myoplasmic Mg^{2+} inhibits or possibly blocks the channel. Some of these effects are illustrated by the single-channel recordings in Figure 3, as well as activation by a non-hydrolysable derivative of ATP. The physiological significance of this is unclear, like the mild activating effect of $Ins(1,4,5)P_3$ (see references 7 and 10). In addition, many ryanodine-sensitive Ca^{2+}-release channels are activated by caffeine, inhibited by the polycationic dye Ruthenium Red, and blocked by local anaesthetics such as procaine and lignocaine. Finally, despite the accepted nomenclature, these channels are also highly permeable to *monovalent* cations[9,11] (Figure 3).

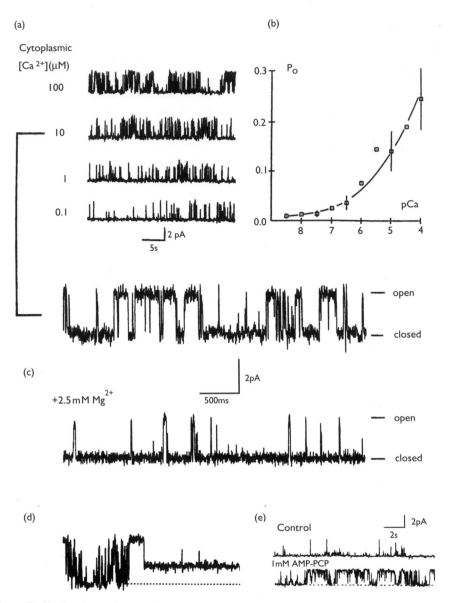

Figure 3. Single-channel recordings
Data from Ca^{2+}-release channels exposed to 50 mM 'luminal' Ca^{2+}, showing unit currents (upwards deflexions) as the ions flow in a lumen-to-myoplasm direction while the bilayer is voltage-clamped at 0 mV. (a–c) Cardiac channels activated by 'myoplasmic' Ca^{2+} and inhibited by (reasonably) physiological levels of myoplasmic Mg^{2+}; bars in (b) represent ±SD of at least six experiments. (d) A brain Ca^{2+}-release channel shortly after exposure to 10 μM 'myoplasmic' ryanodine. Note that the channel is locked into a subconductance state. (e) Another brain channel before and after addition of 1 mM non-hydrolysable ATP-derivative methyleneadenosine 5'-triphosphate (AMP-PCP). (The recording solutions are only crude approximations to native conditions.) P_o represents the probability of a single channel being open. Adapted in part from reference 7 (reproduced from J. Membr. Biol. **111**, 179–189 with permission) and reference 12 (reproduced from J. Gen. Physiol. **95**, 981–1005 with copyright permission from the Rockefeller University Press).

cDNA cloning, and protein sequence and structure

The purification of RyR proteins from muscle SR vesicles (an excellent source, as binding studies suggest that in some skeletal SR fractions the RyR may represent up to about 40% by weight of the membrane protein) facilitated protein sequencing and contributed to the generation of specific antibodies. This enabled both oligonucleotide-directed and expression screening of appropriate cDNA libraries and, in due course, the cloning, sequencing and expression of full-length RyR cDNAs (for example, see reference 13). The availability of several predicted protein sequences relating initially to mammalian skeletal and cardiac muscle SR receptors, and also to a 'brain-type' (type-3) RyR[14,15], has led to intense speculation concerning channel protein structure and possible structure–function relationships. One general interpretation[13] of channel secondary structure is summarized in Figure 4.

While there has been agreement on four putative membrane-spanning α-helical regions close to the C-terminus, some workers have argued for the presence of many additional membrane-spanning segments, often quite far-removed from this region. Also, the significance of many 'consensus' sequences (e.g. those for ligand binding) is disputed. For example, the putative Ca^{2+}-binding EF-hands (helix–loop–helix motifs) noted in the rabbit skeletal RyR primary sequence[13] (Figure 4) are not obvious in the cardiac RyR sequence. In addition, there has been little agreement over the possible assignment of nucleotide-binding (Gly–Xaa–Gly–Xaa–Xaa–Gly) motifs, partly because these appear to be in regions of predicted secondary structure atypical of high-affinity ATP-binding sites. Paradoxically this may be significant, as nucleotides are known to be effective only at relatively high concentrations (0.1–1 mM). This illustrates the difficulties involved in making interpretations from a protein's primary sequence alone, in the absence of a crystal structure (although it is likely that such a large protein would be difficult to 'solve' even if crystallized). Recent attempts to answer many of the important questions related to membrane topology and structure–function relationships will be discussed later.

Meanwhile, image reconstruction techniques, particularly following cryo-electron microscopy[16], have offered an important insight into channel conformation, albeit only on a 3 nm scale. The findings are entirely consistent with a basic model of four subunits arranged around a central pore, but so far the resolution is too low to distinguish between detailed models of membrane topology. The cytoplasmic regions of the protein have a remarkably 'open' organization, and appear to include a 'plug-like' region which may move into and out of the vestibule leading to the cytoplasmic opening of the central pore. This may be highly relevant to the mechanism of EC-coupling.

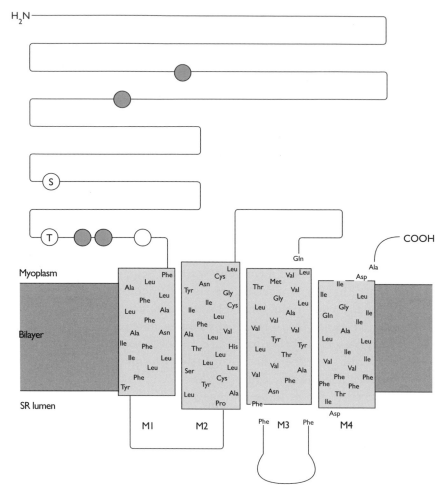

Figure 4. Outline structure of a single rabbit skeletal SR RyR protein subunit
More than 95% of the 5037-residue protein, and both the N- and C-termini, are cytoplasmic. This model[13] assigns four putative transmembrane α-helices (M1–M4, not drawn to scale) with two relatively short intraluminal loops. Several 'consensus' sequences are shown: the filled circles represent putative adenine nucleotide-binding sites, and the clear circle is a region containing putative Ca^{2+}-binding EF-hands. The approximate positions of potentially phosphorylatable serine (S) and threonine (T) residues are also shown. The complete RyR is a homotetramer in which the subunits surround a central water-filled pore.

EC-coupling

Structural analysis of purified skeletal SR receptor proteins[8], and comparison with EM studies of native muscle tissues, have strongly supported the suggestion that the RyR protein contributes to the so-called 'foot process' which appears to bridge the gap between the transverse (t-) tubules of the sarcolemma and the SR membrane. Figure 5 presents a simplified interpretation of this 'junctional' region. In skeletal muscle membrane, depolarization-induced movements of charged protein domains in t-tubular

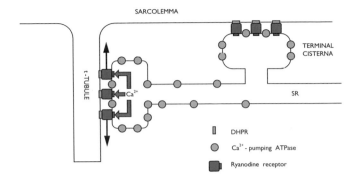

Figure 5. The physical basis of EC-coupling in skeletal muscle
A general outline of the proposed relationship between SR membrane RyR tetramers and DHPR/surface membrane Ca^{2+} channels. The presence of large numbers of RyRs and other proteins in the 'terminal cisternae' of the SR allows the 'heavy' and 'light' SR membrane vesicles to be separated for subsequent flux studies, protein purification and reconstitution. The abundant Ca^{2+}-pumping ATPase is shown, but not the K^+, Cl^- and other channels also known to be present in the SR membrane, nor the t-tubule Na^+ and K^+ channels. Abbreviation: DHPR, dihydropyridine receptor.

membrane proteins, specifically the dihydropyridine receptor (DHPR), may lead to reciprocal movements in charged, capacitatively coupled regions of the RyR in the terminal SR membrane. Note that this is consistent with 'voltage-dependent' gating in the RyR. Some DHPRs are also functional Ca^{2+} channels and, therefore, contribute to Ca^{2+}-induced Ca^{2+} release. In addition, it should be noted that triadin, dystrophin and other cytoskeletal proteins are likely to be very important in the generation, maintenance and overall function of these inter-membrane junctions. One outstanding problem is that the egress of intra-SR Ca^{2+} through a central pore in the RyR would now be obstructed, unless the ions could somehow move laterally through the walls of the foot process. The previously described reconstructions[16] suggest that the foot process of the RyR resembles a loosely woven wicker basket which might allow Ca^{2+} to exit from the 'sides' as well as from the 'mouth' of the channel vestibule. This fulfils a major requirement of this simple model. An outward Ca^{2+} current of some 0.01 pA (which approximates that occurring *in vivo*, assuming intra-SR $[Ca^{2+}]$ is approximately 100 µM and the channels behave the same *in vivo* as *in vitro*) implies the net passage from SR to myoplasm of ~30000 divalent or ~60000 monovalent ions per second, so the 'mesh' cannot pose a significant barrier to ion diffusion. In fact, very much higher flux rates are supported *in vitro*, and single-channel conductances for monovalent cations only saturate at 0.5–1 nS, or 0.5–1 nA per V, well beyond the theoretical maximum for a pore limited by a conventional mouth/vestibule region. (Of course, this does not explain how ions can be admitted equally rapidly from within the limited capture radius of the 'conventional' luminal entrance.)

The less-marked voltage-dependence of the type-2 RyR in cardiac SR, with more reliance on Ca^{2+}-dependent activation, is consistent with the comparatively poorly defined physical association of the cardiac RyR 'foot process' with the sarcolemmal membrane, and the important role of external Ca^{2+} in cardiac EC-coupling. In fact, the large inward Ca^{2+} current during the 'plateau' phase of the cardiac muscle action potential may briefly elevate local cytoplasmic $[Ca^{2+}]$ to 10 μM or more, likely to be enough to maximally activate SR Ca^{2+} release and reduce the need for capacitative coupling between the RyR and proteins in the surface membrane. A more precise description of both cardiac and skeletal EC-coupling now awaits physiological dissection of the modulatory roles of cytoplasmic pH, $[Mg^{2+}]$, [ATP] and [ADP], and many other ligands and accessory proteins, as well as an appreciation of the exact role of channel phosphorylation and other factors.

The Ins(1,4,5)P_3 receptor

Work from many laboratories over the last 10 years has provided very extensive support for a generalized second-messenger role for Ins(1,4,5)P_3. A receptor (InsP_3R) was first purified from mammalian cerebellum[17] and its cDNA was cloned and sequenced[18]. The protein was localized to the ER of cerebellar cells, and purified InsP_3Rs from brain (and smooth muscle) were subsequently shown to mediate appropriate Ca^{2+} fluxes in reconstituted liposomal systems. Functional ion channels were incorporated into planar lipid bilayers from native membrane vesicles and also as purified proteins[19]. However, their single-channel current amplitudes were only 10–20% of those obtained from the RyR under similar conditions, and currents below about 1 pA are often poorly resolved in bilayers because of the unavoidable electrical noise associated with a relatively large membrane area. Nevertheless, a combination of single-channel recording and macroscopic flux studies has shown that the InsP_3R, like the ryanodine-sensitive Ca^{2+}-release channel, is relatively poorly selective for Ca^{2+} (e.g. the permeability ratio of Ca^{2+} to $Tris^+$ was approx. 6; see reference 19), and also conducts monovalent cations, although not as freely as the RyR. Like the RyR, its activity is also modulated by pH, Ca^{2+} and ATP, but it is insensitive to caffeine and specifically inhibited by heparin.

Single Ins(1,4,5)P_3-activated channels show a 'bell-shaped' probability of being open in response to $[Ca^{2+}]$[20], and increasing intraluminal $[Ca^{2+}]$ increases the affinity of the receptor for Ins(1,4,5)P_3[21]. Ins(1,4,5)P_3 can also inactivate the channel, probably acting in concert with cytoplasmic Ca^{2+} (see reference 22). Taken together, these features may help to explain some of the remarkable kinetics of Ins(1,4,5)P_3-induced Ca^{2+} release. For example, release is often 'quantal' in nature, and discrete stores may release all their Ca^{2+} in an all-or-none fashion as cytoplasmic $[Ins(1,4,5)P_3]$ is gradually increased. This is seen most impressively as the eruption of 'hot spots' close to ER membranes in Ca^{2+}-imaging experiments. Some of these phenomena are described in more detail by Berridge[23], and probably contribute to the generation of Ca^{2+} waves,

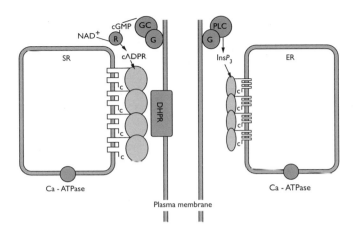

Figure 6. Comparison of Ins(1,4,5)P_3-activated and ryanodine-sensitive Ca^{2+}-release channels

The RyR protein is capacitatively coupled to the DHPR in skeletal muscle, but not in other tissues such as heart and brain, where signalling by cADP-ribose[25] may take place in addition to channel activation by cytoplasmic Ca^{2+}. The process of channel activation may involve lifting a protein 'plug' embedded in the cytoplasmic entrance to the pore, allowing Ca^{2+} to escape through a 'mesh-like' foot process. Activation of the InsP_3R occurs via generation of the messenger Ins(1,4,5)P_3. Little is known about the physical mechanism of the channel, although it is likely to resemble the RyR in important respects. The Ca^{2+} channel subunits have been 'spread out' from their normal cauliflower-like assemblies to show the transmembrane segments more clearly. Abbreviations used: G, receptor or NO-coupled GTP-binding proteins coupled to guanylate cyclase (GC) or phospholipase C (PLC); R, ADP-ribosyl cyclase; DHPR, dihydropyridine receptor.

or oscillations in free [Ca^{2+}], which characterize many cells and their responses to stimulation. Many complicated mathematical models have been developed for such events, but quite simple explanations may account for some of them. For example, microsomal [Ca^{2+}] in vascular smooth muscle cells was recently monitored by taking advantage of the previously mentioned sequestration of fluorescent Ca^{2+} dyes into cellular organelles[24]. Quantal release and incremental detection may simply reflect the relative density of (in this case essentially similar) InsP_3Rs located in the membranes of heterogeneous stores. Finally, interactions between multiple ligands, and the presence of different InsP_3R isoforms whose individual properties [e.g. sensitivity to Ins(1,4,5)P_3 and modulation by phosphorylation] may differ quite substantially, suggests that the mechanisms underlying intracellular Ca^{2+}-mobilization by Ins(1,4,5)P_3 will continue to demonstrate remarkable versatility and complexity.

The physical and chemical properties of InsP_3R proteins show striking similarities to the RyR (Figure 6). In particular, cross-linking studies have shown that InsP_3Rs isolated from mouse cerebellum are formed from four high-molecular-mass homo-oligomers (in this case, subunits of ~320 kDa[19]) containing binding sites for calmodulin and ATP. Secondary structure predictions indicate membrane-spanning α-helices towards the C-terminus of the protein, like the RyR. However, the InsP_3R is definitely glycosylated, as revealed during its purification[17], suggesting a more significant intraluminal representation. This is discussed in more detail in due course. InsP_3Rs can be phosphorylated by cAMP-dependent and other kinases, producing different effects depending on the source or cell type[23]. Some of the structural and functional properties of RyR and InsP_3R proteins are compared in Table 1.

Table 1. Comparison of ryanodine-sensitive and Ins(1,4,5)P_3-activated Ca^{2+}-release channels

Property	RyR[a]	InsP_3R[b]
Molecular mass (deduced from cDNA)	~ 565 kDa[13] 552 kDa[15]	~ 320 kDa[18]
Subunit structure	Homotetrameric[8]	Homotetrameric[19]
Glycosylation	?No	Yes[17]
Transmembrane segments	at least 4[13]	?6[18]
Conductance (50 mM 'luminal' [Ca^{2+}])	100 pS[4,5–7]	10–20 pS[19]
Activators	Voltage Ca^{2+} Ins(1,4,5)P_3[7,10] Adenine nucleotides Caffeine cADP-ribose[25]	Ca^{2+} Ins(1,4,5)P_3 Adenine nucleotides (variably)
Inhibitors	Ryanodine[c] Ca^{2+} [20d] Ruthenium Red	Heparin Ca^{2+} [20] Ins(1,4,5)P_3[22]

References are only provided for the less commonly used activators and inhibitors.

[a]Reviewed in reference 11.

[b]Reviewed in reference 23.

[c]Ryanodine has complex actions on both channel gating and conductance, depending on its concentration (for example, see Figure 3); very low concentrations (~1–10 nM) may simply activate the channel without inducing a substrate.

[d]Inhibition by cytoplasmic and/or luminal Ca^{2+} is not a robust phenomenon for all channels, and appears to depend on many poorly understood factors.

Gene families for intracellular Ca²⁺ channels

Genetic diversity

cDNA cloning studies have shown that each isoform (types 1–3) of the RyR channel is the product of a distinct gene (RYR1–3). A further novel mRNA transcript appears to originate from an alternative transcription initiation site situated towards the 3′-terminal region of RYR1[26], although it is not yet known whether this gives rise to a functional protein product. Some non-mammalian species have more than one 'skeletal muscle-type' RyR, but it is not yet clear whether these result from alternative mRNA splicing or expression from different genes. $InsP_3Rs$ include the protein products of at least four distinct genes[27], complicated by extensive alternative mRNA splicing (outlined in principle in reference 23). In fact, variations in mRNA processing may greatly increase the number of protein isoforms available for each species of intracellular Ca²⁺ channel and, in addition, it is likely that further $InsP_3R$ and RyR genes will be identified. Much more information will undoubtedly come from the continued application of PCR-based homology screening. This has become a remarkably productive tool for cloning multiple members of ion channel gene families[28], using as a template cDNA reverse-transcribed from cell or tissue RNA (Figure 7).

Figure 7. Design of PCR probes for cloning related cDNAs
A PCR strategy to clone multiple $InsP_3R$ cDNAs based on data in references 18 and 27. (a) A partial protein sequence for the longer splice variant of rat Ins(1,4,5)$P_3$1, showing possible primer sites and relevant cDNA sequences. In this example, only non-degenerate PCR primers have been shown in (b).

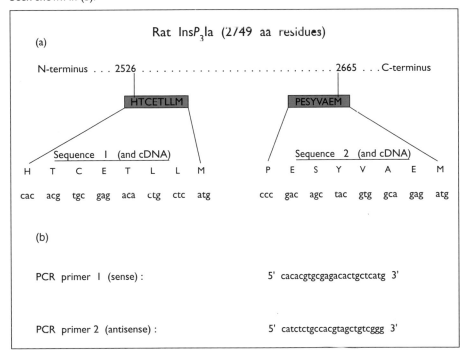

When the same isoforms from different species are compared, RyRs show marked sequence similarities (e.g. >90% identity for rabbit and human type-1). This contrasts with the much reduced similarities observed for different RyR isoforms in the same species (e.g. only 66% identity between rabbit skeletal and cardiac RyRs), consistent with a prototypical gene which has undergone duplication followed by marked divergence. InsP_3Rs and RyRs show similarities to each other but not to any other channel proteins, with the exception of some identical residues in the putative pore-forming regions of both the RyR and the nicotinic acetylcholine receptor[13], and may have diverged from a common ancestral gene as multicellular organisms evolved over 600 million years ago[29]. Finally, the most marked sequence similarities always occur towards the C-terminus of the two families of proteins, consistent with the idea that this part of the sequence makes a substantial contribution to a basic pore-forming region (Figure 6) which has evolved least over time.

Cell- and tissue-specific expression

The tissue- and cell-specific expression of RyR and InsP_3R proteins is currently under very active investigation (for a review, see reference 30). As already mentioned, mammalian skeletal and cardiac muscle contain RyRs type-1 and type-2, respectively, and InsP_3Rs are barely represented at all in these tissues. However, InsP_3RI appears to be present in the cardiac conducting system, and Ins(1,4,5)P_3 activates a frog skeletal muscle SR Ca^{2+} channel. Despite the fact that the first single-channel recordings from InsP_3Rs were made from reconstituted vascular smooth muscle SR, the identity and distribution of Ca^{2+}-release channels in smooth muscle are very poorly characterized compared with skeletal and cardiac muscle. This results in part from the complex anatomy of the relevant endomembrane (ER/SR) system. Careful studies (for example, see reference 6) have shown that some enteric and vascular smooth muscle SR membranes do contain high-affinity binding sites for ryanodine, and the former also contain functional ryanodine-sensitive channels[6].

Mammalian brain contains most of the known RyR and InsP_3R isoforms. Immunoblotting and immunohistochemical studies have localized the mouse type-1 RyR protein to the cerebellum, probably exclusively to cerebellar Purkinje cells, with more widespread expression of the type-2 protein throughout several brain regions (quantitatively, the 'cardiac' type-2 protein is certainly a major 'brain' isoform[10]). In addition, although mRNA levels may not necessarily be correlated with the amount of protein, Northern analysis of the transcription of the RYR3 gene showed strong hybridization of specific cDNA probes to RNA isolated from rabbit corpus striatum, thalamus and hippocampus, with lower levels of expression in pons and medulla, and lowest levels of all in cerebral cortex and cerebellum[15]. *In situ* mRNA hybridization has been used to localize alternatively spliced InsP_3R mRNA transcripts in rat and mouse brain regions, and one important finding has been that different

Table 2. Putative mammalian tissue contents of major InsP$_3$R and RyR isoforms

Tissue	InsP$_3$R	RyR
Skeletal muscle	?Absent in mammalian	**type-1** (?some type-3)
Cardiac muscle	InsP$_3$RI (conducting system)	**type-2** (?some type-1)
Vascular smooth muscle	Various	**types-3 and -2**
Enteric smooth muscle	Various	**types-3** and -2
Brain	InsP$_3$RIa, Ib, II, III, IV	**types-2**, -3 and -1 (minor, cerebellum)
(Mlnk lung) epithelium	—	type-3
Liver	InsP$_3$RIa, Ib	type-3
T-cells	InsP$_3$RIa, Ib	type-3

Bold type indicates major isoforms (where known), and the only splice variants shown are for InsP$_3$RI. Data from a variety of sources, not always based on detection of actual protein product.

InsP$_3$R isoforms may be present in the same cell[31]. This marked level of expressional diversity may well be amplified by the presence of hetero-oligomers. There is the added complication of regional diversity. For example, it is already known that InsP$_3$Rs can be present in nuclear as well as ER membranes in some cells. Some of the rapidly expanding data on the tissue-specific localization of InsP$_3$Rs and RyRs are summarized in Table 2.

Membrane topology and protein structure–function relationships

The InsP$_3$R

Monoclonal and site-directed antibodies have been used in immuno-electron microscopy studies to determine the location of the C-terminus of the protein, and to locate glycosylation sites. An early model[18] of the InsP$_3$R, which assigned six transmembrane helices towards the C-terminus, has recently been supported by a study which combined the use of site-directed mutagenesis, lectin affinity chromatography and site-directed antibodies[32]. At least three residues were glycosylated, and must, therefore, be exposed to the lumen of the ER. These were all sited in regions consistent with the six-helix model. The location of the cytoplasmic Ins(1,4,5)P_3-binding site has also been of interest. In earlier work, both the full-length protein and a shortened version lacking the 418 N-terminal residues were successfully expressed in COS cells. Ins(1,4,5)P_3 binding[33] was abolished in the truncated version. However, monoclonal antibodies recognizing the C-terminus have helped to establish that this region is also cytoplasmic, and it is also involved in both Ins(1,4,5)P_3-binding and channel activation[34]. This suggests that very widely separated regions in the protein's primary sequence may interact closely in the native three-dimensional structure. Finally, at least 50% of the mass of the protein is disposed in the cytoplasm between the N-terminal component of the

Ins(1,4,5)P_3-binding region and the first transmembrane helix. It has been suggested that this domain may contain the important modulatory sites for phosphorylation and Ca^{2+} and adenine nucleotide binding sites, which seem to vary in different isoforms. This outline '3-domain' structure for the InsP_3R family is supported by alignments of rat InsP_3Rs, which showed marked similarities in the N-terminal 'ligand-binding' and C-terminal 'Ca^{2+} channel' domains, with more differences in the 'coupling' domain.

The RyR

More simply represented as two main functional domains, comprising the cytoplasmically disposed multiple-ligand-binding foot-process (at least 95% of the mass of the protein) and the relatively small membrane-associated domain[11] (Figure 4), the RyR also appears to have both its N- and C-termini exposed to the cytoplasm, again consistent with an even number of transmembrane segments. Considerable efforts are currently being directed towards the identification of specific sites for ligand binding and post-translational modification. For example, phosphorylation by ^{32}P followed by protein cleavage and partial sequencing has shown that protein kinase A, protein kinase G and Ca^{2+}/calmodulin-dependent protein kinase (EC 2.7.1.123) all phosphorylate a specific serine residue in the rabbit skeletal muscle RyR[35] (not, in fact, in a consensus sequence, although Ca^{2+}/calmodulin-dependent protein kinase also phosphorylates several other sites). Nucleotide-binding sites have been investigated by covalently attaching a photoreactive ATP analogue containing ^{32}P (see reference 36). Bacterially expressed fusion proteins containing selected lengths of the RyR have been used in $^{45}Ca^{2+}$ and Ruthenium Red overlay techniques, and to raise site-directed polyclonal antibodies, to delineate critical regions involved in Ca^{2+}-activation (see reference 37, although significant findings have also come from studies of mutant channel genes, as described later). Continued intensive efforts to probe structure–function relationships in both of the main Ca^{2+} channel families will lead to significant advances over the next few years.

Other channels

While InsP_3Rs and RyRs remain by far the best-characterized intracellular Ca^{2+} channels, evidence is accumulating from reconstitution and flux studies for the presence of other Ca^{2+}-permeable ion channels in ER and SR membranes. These include voltage-sensitive cation channels, observed by patch-clamping enlarged pancreatic ER microsomes[38] and by incorporating brain microsomes into planar lipid bilayers[39], and also non-selective channels in muscle SR[40]. In addition, it is known that SR[2,4] and ER[7,41] membranes also contain other anion and cation channels, whose role in modulating Ca^{2+} release has hardly been investigated at all at the cellular level. A novel SR Ca^{2+} channel protein has been described in detail[42]. Its molecular mass is 'only' 106 kDa, and it is immunologically distinct from the RyR, despite very similar functional properties; however, its role in EC-coupling is currently unknown. Finally, other channels which have a significant or exclusively plasma membrane representation (e.g. channels activated by $InsP_4$ or by intracellular Ca^{2+} depletion) may also influence intracellular Ca^{2+} mobilization.

New insights into cell regulation and disease

Mutations of the RYR1 gene have been implicated in both human and animal pathologies. Transgenic mice homozygous for the disrupted gene died *in utero* with gross skeletal muscle abnormalities[43], and in man increasingly well-characterized non-lethal mutations of the homologous gene are known to cause certain muscle disorders, including malignant hyperthermia[44,45]. A single amino acid substitution[45] represents the sole alteration in the protein's sequence in malignant hyperthermia-susceptible pigs, but this is sufficient to cause the characteristic increase in Ca^{2+}-sensitivity. Interestingly, this important Ca^{2+}-modulatory site does not correspond to the position of the putative EF-hands, or to Ca^{2+}-binding sites located by other techniques. However, some of these regions may turn out to be associated when the tertiary structure of the protein is known. Many other clinical disorders may be related to inherited or sporadic mutations in RYR2 or 3, to mutations in members of the $InsP_3R$ gene family, or to inherited or acquired abnormalities in the expression of specific genes. For example, heart muscle failure has been associated (though not causally) with abnormal RYR2 expression in some patients[46]. No specific pathology has yet been described for the $InsP_3R$, although lithium, which modifies $InsP_3$ metabolism, is often highly effective in treating bipolar disorder. The widespread expression of $InsP_3Rs$ increases the likelihood of their involvement in specific diseases. For example, in the brain, where cellular Ca^{2+}-handling is so important in the control of neuronal metabolism and electrical activity, and where many intracellular Ca^{2+} channels are in fact represented, it is conceivable that specific abnormalities might contribute to epilepsy or to the extent of Ca^{2+}-dependent brain damage after cerebral ischaemia. A sustained rise in cell $[Ca^{2+}]$ is known to be associated with cell death, and the remodelling and loss of synaptic connections are important in brain development and in certain dementias. In this connection, it may be noted that the expression of the type-3 RyR is growth factor-regulated[14].

Although $InsP_3Rs$ and RyRs are already known to be modulated by many cytoplasmic components, there are good reasons to suppose that even more activators and inhibitors will be discovered. In particular, it is not clear that Ca^{2+}-induced Ca^{2+} release is the sole function of RyRs in brain and other non-contractile tissues. The presence of as yet unidentified physiological ligands would provide a specific role for many ryanodine-sensitive channels. The discovery that cyclic ADP-ribose (cADPR) acts on type-2 RyRs to mobilize intracellular Ca^{2+} (see reference 47), together with the characterization of appropriate cellular machinery[25], has provided just such a sought-after example. The ultimate significance of this by no means general effect, and whether cADPR acts directly on the RyR, are still open questions. In addition, while it is recognized that calmodulin may bind to the RyR as an accessory protein, the discovery that each subunit of the RyR is non-covalently but tightly associated with a relatively small (12 kDa) immunosuppressant drug-binding protein (FK506-binding protein, FKBP)[48] is also highly significant. In retrospect, it is easy to appreciate how a small protein could have been overlooked after SDS/PAGE of the purified RyR, given that loading conditions and acrylamide concentrations were usually optimized for the detection of a much larger (apparent molecular mass 400–500 kDa) protein. Co-expression and reconstitution studies are beginning to suggest that FKBP may have important effects on channel function. The FKBP/FK506 complex inhibits the Ca^{2+}-activated protein phosphatase calcineurin, but the cellular role of FKBP itself has not yet been characterized.

Concluding remarks and future research

The recent discovery and molecular and functional characterization of ryanodine-sensitive and $Ins(1,4,5)P_3$-activated Ca^{2+}-release channels have been very significant developments in research on intracellular Ca^{2+} signalling over the last few years. Although this essay has concentrated on mammalian tissues, at least one member of these related superfamilies of ion channels seems to be present in every eukaryotic cell, and many questions concerning their functions, molecular mechanisms and transcriptional regulation remain unanswered. For example, what exactly is the role of Ca^{2+}-induced Ca^{2+} release in non-contractile cells? How can we study individual Ca^{2+}-release channel isoforms when cells often contain many different channels? And how can we relate channel structure to function in the absence of a crystal structure? Also, how is the expression of different channels regulated in a given cell or tissue, and just how widely implicated are they in disease?

I am grateful to many teachers, students and colleagues for stimulating discussions on Ca^{2+}-signalling, and apologize for the limited selection of references which has inevitably led to the omission of many important original contributions. The author's work was supported by the BHF, MRC and Wellcome Trust.

References

References 1, 11, 23, 28 and 45 are particularly recommended for further reading.

1.   Pozzan, T., Rizzuto, R., Volpe, P. & Meldolesi, J. (1994) Molecular and cellular physiology of intra-cellular calcium stores. *Physiol. Rev.* **74**, 595–636

2.   Miller, C., ed. (1986) *Ion Channel Reconstitution*, Plenum Press, New York

3. Tsien, R.Y. (1980) New calcium indicators and buffers with high selectivity against magnesium and protons: design, synthesis, and properties of prototype structures. *Biochemistry* **19**, 2396–2404

4. Smith, S.J., Coronado, R. & Meissner, G. (1985) Sarcoplasmic reticulum contains adenine-nucleotide-activated calcium channels. *Nature (London)* **316**, 446–449

5. Herrmann-Frank, A., Darling, E. & Meissner, G. (1991) Functional characterization of the Ca^{2+}-gated Ca^{2+}-release channel of vascular smooth muscle sarcoplasmic reticulum. *Pfluegers Arch.* **418**, 353–359

6. Xu, L., Lai, F.A., Cohn, A. et al. (1994) Evidence for a Ca^{2+}-gated ryanodine-sensitive Ca^{2+}-release channel in visceral smooth muscle. *Proc. Natl. Acad. Sci. U.S.A.* **91**, 3294–3298

7. Ashley, R.H. (1989) Activation and conductance properties of ryanodine-sensitive calcium channels from brain microsomal membranes incorporated into planar lipid bilayers. *J. Membr. Biol.* **111**, 179–189

8. Lai, F.A., Erickson, P.H., Rousseau, E., Liu, Q. & Meissner, G. (1988) Purification and reconstitution of the calcium-release channel from skeletal muscle. *Nature (London)* **331**, 315–319

9. Smith, J.S., Imagawa, T., Ma, J., Fill, M., Campbell, K.P. & Coronado, R. (1988) Purified ryanodine receptor from rabbit skeletal muscle is the calcium-release channel of sarcoplasmic reticulum. *J. Gen. Physiol.* **92**, 1–26

10. Lai, F.A., Dent, M., Wickenden, C. et al. (1992) Expression of a cardiac Ca^{2+}-release channel isoform in mammalian brain. *Biochem. J.* **288**, 553–564

11. Coronado, R., Morrissette, J., Sukhareva, M. & Vaughan, D.M. (1994) Structure and function of ryanodine receptors. *Am. J. Physiol.* **266**, C1485–C1504

12. Ashley, R.H. & Williams, A.J. (1990) Divalent cation activation and inhibition of single calcium-release channels from sheep cardiac sarcoplasmic reticulum. *J. Gen. Physiol.* **95**, 981–1005

13. Takeshima, H., Nishimura, S., Matsumoto, T. et al. (1989) Primary structure and expression from complementary DNA of skeletal muscle ryanodine receptor. *Nature (London)* **339**, 439–445

14. Giannini, G., Clementi, E., Ceci, R., Marziali, G. & Sorrentino, V. (1992) Expression of a ryanodine receptor-Ca^{2+} channel that is regulated by TGF-β. *Science* **257**, 91–94

15. Hakamata, Y., Nakai, J., Takeshima, H. & Imoto, K. (1992) Primary structure and distribution of a novel ryanodine receptor/calcium-release channel from rabbit brain. *FEBS Lett.* **312**, 229–235

16. Radermacher, M., Rao, V., Grassucci, R. et al. (1994) Cryo-electron microscopy and three-dimensional reconstruction of the calcium-release channel/ryanodine receptor from skeletal muscle. *J. Cell Biol.* **127**, 411–243

17. Supattapone, S., Worley, P.F., Baraban, J.M. & Synder, S.H. (1988) Solubilization, purification, and functional reconstitution of an inositol trisphosphate receptor. *J. Biol. Chem.* **263**, 1530–1534

18. Furuichi, T., Yoshikawa, S., Miyawaki, A., Wada, K., Maeda, N. & Mikoshiba, K. (1989) Primary structure and functional expression of the inositol 1,4,5,-trisphosphate-binding protein P400. *Nature (London)* **342**, 32–38

19. Maeda, N., Kawasaki, T., Nakade, S. et al. (1991) Structural and functional characterization of inositol 1,4,5-trisphosphate receptor channel from mouse cerebellum. *J. Biol. Chem.* **266**, 1109–1116

20. Bezprozvanny, I., Watras, J. & Ehrlich, B.E. (1991) Bell-shaped calcium-response curves of $Ins(1,4,5)P_3$- and calcium-gated channels from endoplasmic reticulum of cerebellum. *Nature (London)* **351**, 751–754

21. Oldershaw, K. & Taylor, C.W. (1993) Luminal Ca^{2+} increases the affinity of inositol 1,4,5-trisphosphate for its receptor. *Biochem. J.* **292**, 631–633

22. Hajnoczky, G. & Thomas, A.P. (1994) The inositol trisphosphate calcium channel is inactivated by inositol trisphosphate. *Nature (London)* **370**, 474–477

23. Berridge, M.J. (1993) Inositol trisphosphate and calcium signalling. Nature (London) **361**, 315–325

24. Hirose, K. & Iino, M. (1994) Heterogeneity of channel density in inositol-1,4,5-trisphosphate-sensitive Ca^{2+} stores. *Nature (London)* **372**, 791–794

25. Galione, A., White, A., Willmott, N., Turner, M., Potter, B.V.L. & Watson, S.P. (1993) cGMP mobilizes intracellular Ca^{2+} in sea-urchin eggs by stimulating cyclic ADP-ribose synthesis. *Nature (London)* **365**, 456–457

26. Takeshima, H., Nishimura, S., Nishi, M., Ikeda, M. & Sugimoto, T. (1993) A brain-specific transcript from the 3'-terminal region of the skeletal muscle ryanodine receptor gene. *FEBS Lett.* **322**, 105–110

27. Ross, C.A., Danoff, S.K., Schell, M.J., Snyder, S.H. & Ullrich, A. (1992) Three additional inositol-1,4,5 trisphosphate receptors: molecular cloning and differential localization in brain and peripheral tissues. *Proc. Natl. Acad. Sci. U.S.A.* **89**, 4265–4269

28. Vega-Saenz de Miera, E.C. & Lin, J. (1992) Cloning of ion channel gene families using the polymerase chain reaction. *Methods Enzymol.* **207**, 613–619

29. Hille, B. (ed.) (1992) *Ionic Channels of Excitable Membranes,* 2nd edn, p. 542, Sinauer, MA

30. Sorrentino, V. & Volpe, P. (1993) Ryanodine receptors: how many, where and why? *Trends Pharmacol. Sci.* **14**, 98–103

31. Ross, C.A., Danoff, S.K., Schell, M.J., Snyder, S.H. & Ullrich, A. (1992) Three additional inositol 1,4,5-trisphosphate receptors: molecular cloning and differential localization in brain and peripheral tissues. *Proc. Natl. Acad. Sci. U.S.A.* **89**, 4265–4269

32. Michikawa, T., Hamanaka, H., Otsu, H. *et al.* (1994) Transmembrane topology and sites of N-glycosylation of inositol 1,4,5-trisphosphate receptor. *J. Biol. Chem.* **269**, 9184–9189

33. Mignery, G.A., Newton, C.L., Archer, B.T. & Sudhof, T.C. (1990) Structure and expression of the rat inositol 1,4,5-trisphosphate receptor. *J. Biol. Chem.* **265**, 12679–12685

34. Nakade, S., Maeda, N. & Mikoshiba, K. (1991) Involvement of the C-terminus of the inositol 1,4,5-trisphosphate receptor in Ca^{2+} release analysed using region-specific monoclonal antibodies. *Biochem. J.* **277**, 125–131

35. Suko, J., Maurer-Fogy, I., Plank, B. et al. (1993) Phosphorylation of serine 2843 in ryanodine receptor-calcium-release channel of skeletal muscle by cAMP-, cGMP- and CaM-dependent protein kinase. *Biochim. Biophys. Acta* **1175**, 193–206

36. Zarka, A. & Shoshan-Barmatz, V. (1993) Characterization and photoaffinity-labelling of the ATP binding site of the ryanodine receptor from skeletal muscle. *Eur. J. Biochem.* **213**, 147–154

37. Treves, S., Chiozzi, P. & Zorzato, F. (1993) Identification of the domain recognized by anti-(ryanodine receptor) antibodies which affect Ca^{2+}-induced Ca^{2+} release. *Biochem. J.* **291**, 757–763

38. Schmid, A., Dehlinger-Kremer, M., Schulz, I. & Gögelein, H. (1990) Voltage-dependent $InsP_3$-insensitive calcium channels in membranes of pancreatic endoplasmic reticulum vesicles. *Nature (London)* **346**, 374–376

39. Martin, C. & Ashley, R.H. (1993) Reconstitution of a voltage-activated calcium conducting cation channel from brain microsomes. *Cell Calcium* **14**, 427–438

40. Sukhareva, M., Morrissette, J. & Coronado, R. (1994) Mechanism of chloride-dependent release of Ca^{2+} in the sarcoplasmic reticulum of rabbit skeletal muscle. *Biophys. J.* **67**, 751–765

41. Silvestro, A.M. & Ashley, R.H. (1994) Solubilization, partial purification and functional reconstitution of a sheep brain endoplasmic reticulum anion channel. *Int. J. Biochem. Mol. Biol.* **26**, 1129–1138

42. Hilkert, R., Zaidi, N., Shome, K., Nigam, M., Lagenaur, C. & Salama, G. (1992) Properties of immunoaffinity purified 106-kDa Ca^{2+}-release channels from the skeletal sarcoplasmic reticulum. *Arch. Biochem. Biophys.* **292**, 1–15

43. Takeshima, H., Iino, M., Takekura, H. *et al.* (1994) Excitation–contraction uncoupling and muscular degeneration in mice lacking functional skeletal muscle ryanodine-receptor gene. *Nature (London)* **369**, 556–559

44. Gillian, C.M., Heffron, J.J., LeHane, M., Marks, A. & McCarthy, T.V. (1991) Analysis of expression of the human ryanodine receptor gene in malignant hyperthermia skeletal muscle. *Biochem. Soc. Trans.* **19**, 46S

45. MacLennan, D.H., Otsu, K., Fujii, J. *et al.* (1992) The role of the skeletal muscle ryanodine receptor gene in malignant hyperthermia. *Symp. Soc. Exp. Biol.* **46**, 189–201

46. Brillantes, A.M., Allen, P., Takahashi, T., Izumo, S. & Marks, A.R. (1992) Differences in cardiac calcium-release channel (ryanodine receptor) expression in myocardium from patients with end-stage heart failure caused by ischemic versus dilated cardiomyopathy. *Circ. Res.* **71**, 18–26

47. Meszaros, L.G., Bak, J. & Chu, A. (1993) Cyclic ADP-ribose as an endogenous regulator of the non-skeletal type ryanodine receptor Ca^{2+} channel. *Nature (London)* **364**, 76–78

48. Jayaraman, T., Brillantes, A.M., Timerman, A.P. et al. (1992) FK506-binding protein associated with the calcium-release channel (ryanodine receptor). *J. Biol. Chem.* **267**, 9474–9477

8

Diphtheria toxin-based receptor-specific chimaeric toxins as targeted therapies

Éamonn B. Sweeney and John R. Murphy

Section of Biomolecular Medicine, Evans Memorial Department of Clinical Research and Department of Medicine, Boston University Medical Center Hospital, 88 East Newton Street, Boston, MA 02118, U.S.A.

Introduction

In the last century, based upon observations with dyes, toxins and antibodies, Paul Ehrlich postulated the development of 'magic bullets' for the treatment of human disease. Ehrlich envisioned the use of antibodies conjugated to deadly toxins to specifically eliminate cancer cells. This vision coupled with the technical advances made during the last 20 years has led to the development of clinically viable chimaeric molecules whereby antibodies or hormones act as targeting vehicles for the delivery of radionuclides, drugs or toxins to target cells. The advances which have been made in the development of targeted therapies have been significant and many chimaeric molecules have shown efficacy *in vivo*. This article will outline the basic science which has helped to establish two classes of site-directed drug currently used in Phase I/II clinical trials. Initially, we will briefly introduce the immunotoxins (ITs), a class of chimaeric drug in which toxic molecules are chemically conjugated to antibodies; we will then concentrate on a more in-depth analysis of the fusion toxins in which toxic moieties are fused genetically to a polypeptide hormone or growth factor. To demonstrate the sequence of events in the generation of such a recombinant fusion toxin, we will summarize the genetic construction and characterization of an interleukin-7 (IL-7) receptor targeted fusion protein currently under investigation in this laboratory.

Toxins

ITs and recombinant fusion toxins have been generated with a variety of plant and bacterial toxins. In general, three toxins have been preferentially used in the assembly of chimaeric proteins, largely because of a detailed understanding of their structure–function relationships. Diphtheria toxin (DT) and *Pseudomonas* exotoxin A (PE), both bacterial toxins, and the plant toxin ricin have been used extensively as the toxic elements in either the ITs or the fusion toxins. These all act enzymically to inhibit cellular protein synthesis and are among the deadliest substances known to man. In general, these protein toxins are composed of receptor binding, transmembrane and catalytic domains. In all instances it is the structure–function organization of the toxin that allows for the isolation of the catalytic domains used in the assembly of ITs and the catalytic and transmembrane domains used in the genetic construction of fusion proteins. These chimaeric molecules have the potential to be more effective than standard chemotherapeutic agents: ITs and fusion toxins have been shown (i) to be effective against both dormant and cycling cells, while most chemotherapies are effective only against those which are actively dividing; (ii) to have low adverse effects, while traditional drugs are non-selective in that they target both normal and malignant cell types, resulting in high toxicities; and (iii) to be extremely lethal, thereby requiring only small doses to be therapeutically effective.

Diphtheria toxin

DT is a 535-amino acid, single-chain polypeptide with a deduced molecular mass of 58342 Da. DT contains two disulphide bridges which are formed between Cys^{186}–Cys^{201} and Cys^{461}–Cys^{471} (Figure 1). The 14-amino acid loop subtended by Cys^{186}–Cys^{201} forms a protease-sensitive region which is readily

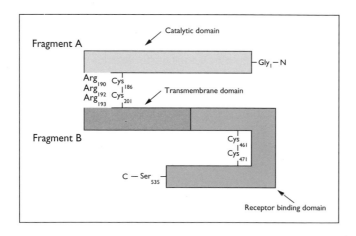

Figure 1.
Schematic diagram of DT
Fragment A (Gly_1-Arg_{193}) contains the catalytic centre for the NAD-dependent ADP-ribosylation of elongation factor 2 (EF-2). Fragment B (Ser_{194}-Ser_{535}) contains the transmembrane domain, as well as the eukaryotic cell receptor-binding domain.

'nicked' by serine proteases. Nicked DT may be separated into an A and a B fragment upon reduction of the disulphide bond under denaturing conditions. The N-terminal fragment A (21 167 Da; 193 amino acids) is the catalytic domain of DT and is active in the ADP-ribosylation of elongation factor 2 (EF-2). The C-terminal fragment B (37 199 Da, 342 amino acids) carries both the transmembrane and receptor-binding domains of the toxin.

Delivery of the catalytic domain (fragment A) of DT to the cytosol of sensitive eukaryotic cells is thought to result from the following steps (Figure 2): (i) toxin binding to its receptor on the cell surface; (ii) internalization of the toxin by receptor-mediated endocytosis; (iii) acidification of the endocytic vesicle, resulting in a conformational change in the transmembrane domain of fragment B; (iv) insertion of the transmembrane domain in the endocytic vesicle membrane, which facilitates delivery of the catalytic domain to the cytosol; culminating in (v) the ADP-ribosylation of EF-2 which leads to the death of the cell. (For a review on DT, see reference 1.)

Figure 2. The expected mechanism of fragment A transfer to a sensitive eukaryotic cell
The DT receptor-binding domain is highlighted in blue and the catalytic fragment A in grey. See text for explanation.

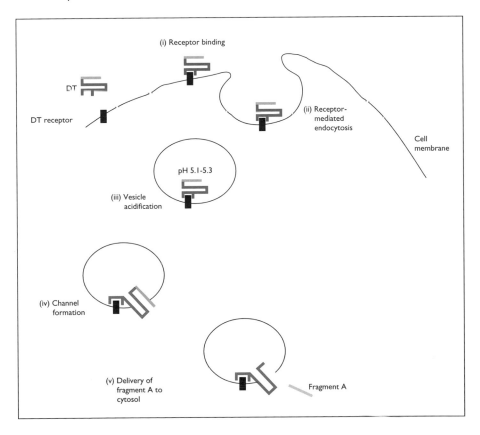

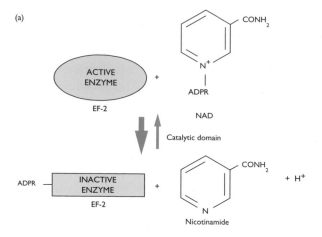

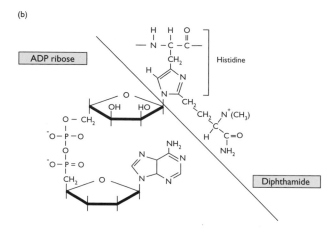

Figure 3. ADP-ribosylation of EF-2 by DT
(a) The catalytic domain of DT catalyses the transfer of the ADP-ribosyl moiety of NAD$^+$ to EF-2 with the release of nicotinamide and a proton. ADP-ribosylated EF-2 thus formed is catalytically inactive and can no longer participate in protein synthesis. In this way, the toxin irreversibly inhibits protein synthesis and kills the host cells. (b) Structure of ADP-ribosylated diphthamide in EF-2.

The ADP-ribosyltransferase (EC 2.4.2.31) reaction catalysed by the catalytic domain, fragment A, is outlined in Figure 3(a). NAD is comprised of an ADP-ribose attached covalently to nicotinamide through a β-N-glycosidic linkage. This high-energy bond supplies the driving force for the ADP-ribosylation of EF-2 by DT. Fragment A catalyses the transfer of the ADP-ribose moiety of NAD$^+$ to a nitrogen atom on the substituted imidazole ring of diphthamide (Figure 3b). All eukaryotic cells, as well as several archae-bacteria, contain diphthamide in their EF-2. Diphthamide is formed by post-translational modification of histidine. Once EF-2 is ADP-ribosylated, it is unable to catalyse the transfer of nascent peptidyl-tRNA from the aminoacyl site to the polypeptidyl site of the eukaryotic ribosome, and, as a result, cellular protein synthesis is inhibited. Yamaizumi *et al.* (1985) have demonstrated that the delivery of a single molecule of the catalytic domain of DT to the eukaryotic cell cytosol is sufficient to kill that cell[2]. (For a review on ADP-ribosylating enzymes, see reference 3.)

Pseudomonas exotoxin A

PE is a protein toxin of 66 kDa secreted by *Pseudomonas aeruginosa*. PE catalyses the same ADP-ribosylation reaction as DT, inactivating EF-2. Structurally, PE and DT are similar in that they have three functional domains: B, binding domain; T, translocation domain; and A, catalytic domain. However, PE is in the opposite structural orientation, with the binding domain located at the N-terminus, the catalytic fragment at the C-terminal end and the transmembrane domain located between the two.

Ricin toxin

Ricin is composed of two functionally distinct polypeptides linked by a disulphide bond. The A chain carries the toxin's catalytic domain while the B chain is responsible for cell-surface binding and subsequent internalization of the A chain into the cell. The A chain enzymically removes adenine A_{4324} from the 60S ribosomal subunit, resulting in a decreased capacity for binding EF-2 and thereby inhibiting protein synthesis and causing cell death. It has been suggested that the delivery of one molecule of ricin A chain to the cytosol is sufficient to kill the cell[4].

Immunotoxins

One of the first examples of an antibody delivering a molecule to its target antigen was described by McLintock and Friedman (1945) and demonstrated the specific localization to cellular target antigens of antibodies coupled to dyes and metals[5]. With the advent of monoclonal antibody (mAb) technology, came the opportunity to develop mAb-based therapies which offered more selectivity. Effective ITs have been constructed in which a mAb has been chemically linked to either holotoxin or the catalytic domain of a toxin. Holotoxin ITs, although usually most potent, dampen the intended specificity of the mAb used; therefore, toxins with modified receptor-binding domains are more frequently used. Since murine mAbs have been generally used in ITs, a common difficulty encountered in human clinical trials has been the development of human anti-mouse antibodies (HAMAs). Recently, the grafting of murine complementarity-determining regions (CDRs) onto human antibody frameworks has been used to minimize the problem of immunogenicity to ITs *in vivo*.

Table 1. Examples of ITs and their target cells

Immunotoxin conjugate	Binding site	Target cells	References
Ricin A–anti-T101	CD5	Leukaemia cells	9
Ricin A–anti-melanoma	220 kDa antigen	Melanoma cells	9
Ricin A–CD4	gp 120	HIV-producing cells	11
PE–CD4	gp 120	HIV-producing cells	12
DT–UCHT1	CD45	Leukaemia cells	13

Abbreviation used: gp, glycoprotein.

The first IT was developed by Moolten and Cooperbrand (1970) by coupling DT to a polyclonal antibody directed against the mumps virus[6]. The first treatment of animals with an IT was also carried out by Moolten et al. (1975) using DT conjugated to a mAb directed against SV-40 transformed sarcoma cells. Importantly, this treatment resulted in a partial to complete regression of some tumours[7]. A few ITs and their target cells are listed in Table 1.

Theoretically, DT is an attractive toxin for IT technology due to its potency; however, in practise ITs assembled with DT fragment A have proven to be either of low potency or inactive. Potent ITs have been assembled with full-length DT, but non-specific toxicities exist due to the presence of the native DT-binding domain. The relative impotency of DT fragment A ITs was explained by Bacha et al.[8] These investigators demonstrated that the hydrophobic region (transmembrane domain) of DT fragment B is required for the efficient translocation of the catalytic domain into the cell. Ricin A-chain-based immunoconjugates were found almost universally to be more potent than those assembled with DT fragment A and, therefore, ricin A chain became the toxophore of choice. (For reviews on ITs, see references 9 and 10.)

Genetically engineered fusion toxins

The joining of the molecular genetics of microbial toxins and cytokine research has allowed for the generation of a new class of cell receptor-specific fusion proteins, namely the fusion toxins. Here, cytokines have been used to replace the native binding moieties of either DT or PE, thereby targeting only those cells which have the specific cytokine receptor. In general, these recombinant fusion proteins offer several advantages over the ITs: (i) the cytokine and toxin are joined by a peptide bond rather than a disulphide bond; (ii) recombinant fusion toxins are expressed as single-chain proteins and have a defined structure, whereas the chemical cross-linking assembly used for the ITs produces a mixture of mAb-toxins where the toxophore may be cross-linked to many different sites on the antibody; (iii) the cytokine/receptor binding is generally of high affinity, whereas mAb binding to its determinant is of intermediate to low affinity; and (iv) cytokine receptors are known to be internalized by receptor-mediated endocytosis which allows for the natural (native) intoxication processes of these toxins to proceed. In contrast, many cell-surface determinants recognized by mAbs are not internalized. In addition, since the fusion toxins are constructed at the genetic level, one can readily conduct structure–function mutation analyses which can optimize cytotoxic potency.

With the determination of the nucleotide sequence of the DT *tox* gene, and the technological advances of oligonucleotide synthesis *in vitro*, came the opportunity for the construction of the first DT-related fusion toxin under Biosafety level 4 containment. Murphy *et al.*[14] chose to link α-MSH (melanocyte-stimulating hormone) to the catalytic and transmembrane domains of DT for the following reasons: (i) the oligonucleotide encoding the primary sequence of α-MSH (13 amino acids) could easily be synthesized *in vitro*; (ii) it was previously shown that the α-MSH binding site was positioned on the C-terminal end of the peptide, thereby allowing genetic fusion at the N-terminus; and (iii) α-MSH is internalized by receptor-mediated endocytosis. The DT-related/α-MSH fusion toxin, DAB_{486}–α-MSH, proved to be highly toxic to cells bearing the α-MSH receptor and not toxic to cells devoid of the receptor. This early work demonstrated that it was possible to replace genetically the binding domain of DT with a polypeptide hormone, and direct DT's cytotoxicity to cells bearing the hormone's receptor.

The IL-2 receptor (IL-2R) is present in large numbers on the surface of several T- and B-cell leukaemias and lymphomas. This receptor would, therefore, be an inviting target for either IT or fusion toxin therapy. Considering that anti-IL-2R antibodies are not readily internalized, a fusion protein consisting of the IL-2 cytokine linked to a toxic molecule should have a better chance of attaining internalization and causing cell death than would an antibody-based IT. In 1987, Williams and co-workers replaced the C-terminal 50 amino acids of DT with the structural gene for IL-2; the resulting fusion toxin, DAB_{486}–IL-2, was found to be remarkably toxic for only those cells bearing the high-affinity IL-2R[15]. The group also demonstrated that the in-frame deletion of 97 amino acids from Thr^{387} to His^{485} of DAB_{486}–IL-2 increased binding and cytotoxicity of the resulting variant, DAB_{389}–IL-2, by approximately 10-fold. The shorter form of the IL-2 fusion toxin inhibited protein synthesis in cells bearing the high-affinity form of the IL-2R by 50% of control at a concentration of approx. $2–3 \times 10^{-12}$ M. In a Phase I clinical trial, DAB_{486}–IL-2 was found to be well-tolerated and showed significant anti-tumour activity even in chemotherapy-resistant patients[16]. Other DT-related fusion toxins have been constructed, such as those targeting the IL-4, IL-6 and CD4 receptors[17–19]. A list of the DT-related fusion toxins which have been generated in this laboratory and some of the cell types they target is shown in Table 2. All of these fusion proteins were shown to have cytotoxic effects on cells bearing their corresponding receptors. (For reviews on DT-related fusion toxins, see references 20 and 21.)

Table 2. A list of DT-related fusion proteins and some of their target cells

Fusion protein	Target cells	Reference
DAB$_{486}$–α-MSH	Melanoma cells	14
DAB$_{486}$–IL-2	Lymphoma/leukaemia cells	15
DAB$_{389}$–EGF	EGF receptor-positive malignancies	22
DAB$_{389}$–IL-4	T-cells	17
DAB$_{389}$–IL-6	Myeloma cells	18
DAB$_{389}$–CD4	HIV-1-infected cells	19

Abbreviations used: EGF, epidermal growth factor; HIV, human immunodeficiency virus.

DAB$_{389}$–IL-7: a new member of the fusion toxin family

Interleukin-7

IL-7 is a 25 kDa glycoprotein originally identified for its ability to sustain the growth of pre-B-cells in long-term bone marrow cultures. IL-7 mRNA was later found in human spleen and thymus, indicating another role for this cytokine in T-cell development. It was subsequently shown that resting fetal and adult thymocytes proliferated in response to IL-7, independently of other known T-cell growth factors, such as IL-2, IL-4 and IL-6. IL-7 also induced the proliferation of mature human T-cells and caused the secretion of cytokines, such as tumour necrosis factor α, from human monocytes. Studies on human haematopoietic malignancies have shown that IL-7 stimulates B and T acute lymphoblastic leukaemia (ALL) cells. Other studies demonstrate the IL-7-driven proliferation of cells derived from patients with chronic lympho-cytic leukaemia (CLL) and the Sezary syndrome of cutaneous T-cell lymphoma. IL-7 was shown to have proliferative effects for malignant cells not only from the lymphoid lineage but to also induce an increase in DNA synthesis in acute myelogenous leukaemia (AML) cells.

Molecular and biochemical characterization of the IL-7 receptor (IL-7R) became possible after cloning the human and murine cDNAs. Goodwin and co-workers isolated cDNA clones for both receptors and expressed them in COS-7 cells[23]. This resulted in cell-surface expression of IL-7R which was capable of binding IL-7 with the same properties as the native receptor. As well as isolating a cDNA for the membrane-bound IL-7R, Goodwin obtained a cDNA encoding a soluble form of the IL-7R. When this cDNA was expressed, it encoded a protein capable of binding IL-7 in solution, which acted as a soluble inhibitor for the ligand. This is similar to the situation with IL-4, which has a receptor in both membrane-bound and soluble forms. Binding of radiolabelled IL-7 to the recombinant IL-7Rs identified high- and low-affinity binding classes (1×10^{10} M^{-1} and 4×10^8 M^{-1} respectively). Analysis of the IL-7R protein sequence identifies it as a member of the cytokine receptor superfamily. IL receptors 2–6 and GM-CSF (granulocyte/

macrophage colony stimulating factor) are all members of this family. Flow cytometry analyses have verified the expression of IL-7Rs on cells of both lymphoid and myelomonocytic origins. T-cell and early B-cell lymphomas were shown to have between 2000 and 4000 receptors per cell. Like the IL-2R, the IL-7R does not have any protein tyrosine kinase domain. Even so, IL-7 (like IL-2, IL-4 and GM-CSF) rapidly induces tyrosine phosphorylation after receptor binding, despite the fact that their receptors themselves are not tyrosine kinases. Therefore, IL-7 has been identified as a member of the group of tyrosine kinase-activating cytokines.

Design and synthesis of the DAB$_{389}$–IL-7 gene

A synthetic allele for human IL-7 was generated to incorporate restriction endonuclease sites that would allow cloning immediately downstream of, and in-frame with, the truncated DT construct DAB$_{389}$, which encodes the transmembrane and catalytic regions of DT. The total oligonucleotide synthesis *in vitro* would also allow a template design based on *Escherichia coli* usage bias. The gene was designed using *E. coli* codon usage bias to favour protein production in the bacterium. The PCR method employed in this work synthesized the gene in two halves; each half consisted of a pair of templates with complementary overhangs which were PCR amplified by flanking 5′ and 3′ primers. Each half was thus generated and annealed together at the unique *Age*1 restriction site in the middle of the gene. This procedure, outlined in Figure 4, generated the full IL-7 gene with flanking restriction sites to allow

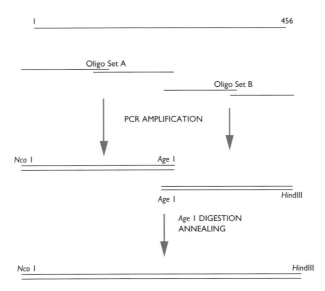

Figure 4. The synthetic IL-7 gene synthesized from four overlapping oligonucleotides spanning the entire length of the gene
The 5′ to 3′ construction of the gene was accomplished by PCR amplification of oligonucleotide sets A and B which yielded two PCR fragments, each containing one half of the IL-7 gene. The full recombinant human (rhIL-7) was generated by annealing both *Age*1-digested PCR products.

for in-frame cloning into DAB_{389}. The resulting gene fusion, DAB_{389}–IL-7, became another member of a family of fusion genes where the structural gene for the binding domain of DT has been replaced with the gene for ligands such as α-MSH, IL-2, IL-4 and IL-6.

DAB_{389}–IL-7 expression and purification

Having constructed the gene for DAB_{389}–IL-7, the fusion protein was expressed in a T7 expression system under the control of an isopropyl β-D-thiogalactoside (IPTG)-inducible T7 promoter. Foreign proteins expressed in this system accumulate in the *E. coli* cytoplasm as insoluble inclusion bodies. Although the proteins in these bodies are not refolded, their insolubility allows for a relatively high degree of purification from *E. coli* proteins. The inclusion bodies are denatured in 5 M guanidine hydrochloride solution and refolded by dialysis (using a refolding protocol described by E.B. Sweeney, L. Cosenza and J.R. Murphy, unpublished work). SDS/PAGE and Western blot analyses using antibodies against DT and IL-7 identified the inclusion body protein as having both DT and IL-7 immunological determinants. The molecular mass of the fusion protein was equal to that of DAB_{389} and IL-7 combined. The percentage of total monomer in the protein preparation was determined from a native PAGE to be approximately 10%, with the remainder of the protein in higher multimeric forms. In this case, only 10% of total

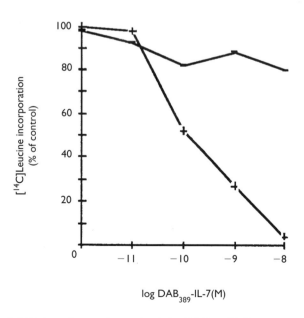

log DAB_{389}-IL-7(M)

Figure 5. The inhibition of protein synthesis by DAB_{389}–IL-7 as measured by the incorporation of ^{14}C-leucine
Data is plotted as the percentage of ^{14}C-leucine incorporation of control cells treated with medium alone. +, IL-7 receptor-positive 2E8 cells; –, IL-7 receptor-negative CHO cells.

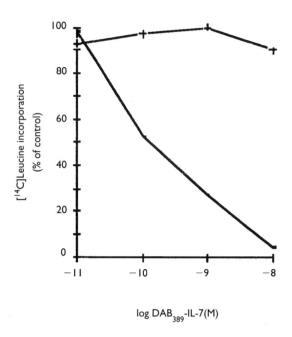

Figure 6. Anti-IL-7 polyclonal antibody-mediated neutralization of DAB$_{389}$–IL-7 activity

2E8 cells were incubated with 10^{-8}–10^{-11} M DAB$_{389}$–IL-7 in the presence and absence of 10 mg/ml IL-7 polyclonal antibody (pAb). Results are reported as the percentage of leucine incorporation of the control (no toxin). +, Toxin-treated 2E8 cells in the presence of 10 μg/ml IL-7 pAb; −, 2E8 cells treated with toxin alone.

protein will be active, since aggregated forms of fusion toxins are inactive (J. vanderSpek, personal communication).

Protein synthesis inhibition in an IL-7R+ cell line

Inclusion body purified DAB$_{389}$–IL-7 was tested for its cytotoxic activity against the IL-7-dependent (IL-7R+) 2E8, and IL-7R negative CHO cell lines. It was demonstrated that this fusion toxin is active only against those cell lines bearing the IL-7R that can be internalized *in vitro*. Moreover, since the addition of excess IL-7 or IL-7 polyclonal antibody to the assay mixture blocks the action of this fusion toxin, it was concluded that the cytotoxic action of DAB$_{389}$–IL-7 is targeted by the IL-7 component of the fusion protein. The toxin inhibited protein synthesis in the 2E8 cells at an IC$_{50}$ of 1×10^{-10} M, whereas it had no effect on protein synthesis at the highest concentration used (10^{-8} M) in the IL-7R negative Chinese hamster ovary (CHO) line (Figure 5). The cytotoxic activity of DAB$_{389}$–IL-7 was neutralized upon the addition of 10 μg/ml of IL-7 polyclonal antibody (Figure 6). We have also generated an adenosine diphosphate ribosyl-transferase-inactive mutant DAB$_{389}$–IL-7 fusion protein, which does not kill IL-7R+ cells but actually induces the IL-7-dependent line to proliferate. This shows that the cytokine-binding component of the fusion toxin is still active in the chimaera. These data indicate that the fusion toxin binds to the IL-7R on target cells through the IL-7 binding domain and that cell killing is due to the enzymic activity of the catalytic domain of DT. The IC_{50} for DAB_{389}–IL-7 reported in this work is of the same order of magnitude as has been reported for other DAB_{389}-related IL fusion toxins. Although IC_{50} is a function of receptor number, the IC_{50} values reported for DAB_{389}–IL-4 and DAB_{389}–IL-6 were 5×10^{-10} M and 2×10^{-11} M, respectively.

Summary

- *The results from the phase I/II studies of the intravenous administration of DAB_{486}–IL-2 to patients with refractory haematological malignancies have now proven in principle the feasibility of fusion toxin therapy in man. Indeed, the cell-surface receptor-specific intoxication of neoplastic cells through the catalytic ADP-ribosylation of EF-2 is the prototype of a new class of biological response modifiers that may be generally applicable.*

- *In those circumstances where either the* de novo *expression or up-regulation of a cell-surface receptor can be associated with human disease [e.g. the up-regulation of the epidermal growth factor (EGF) receptor on breast cancer], it should be possible to construct genetically a DT-related/growth factor fusion protein to produce an experimental biological treatment of that malignancy.*

- *The EGF receptor-targeted fusion toxin DAB_{389}–EGF has within the last year begun human phase I clinical trials. The pre-clinical development of DAB_{389}–IL-7 has begun with the anticipation that this novel fusion toxin will be evaluated in the treatment of the acute leukaemias in which the IL-7R has been shown to be present.*

References

1. Pappenheimer, A.M., Jr. (1977) Diphtheria Toxin. *Annu. Rev. Biochem.* **6**, 69–94
2. Yamaizumi, M., Mekada, E., Uchida, T. & Okada, Y. (1978) One molecule of diphtheria toxin fragment A introduced into a cell can kill the cell. *Cell* **15**, 245–250
3. Hayaishi, O. & Ueda, K. (1977) Poly(ADP-ribose) and ADP-ribosylation of proteins. *Annu. Rev. Biochem.* **46**, 95–116
4. Eiklid, K., Olsnes, S. & Pihl, A. (1980) Entry of lethal doses of abrin, ricin and modeccin into the cytosol of Hela cells. *Exp. Cell Res.* **126**, 321–326
5. McLintock, L.A. & Friedman, M.M. (1945) Utilization of antibody for the localization of metals and dyes in the tissues. *Am. J. Roentgenol. Radium Therapy* **54**, 704–706
6. Moolten, F.L. & Cooperbrand, S.R. (1970) Selective destruction of target cells by diphtheria toxin conjugated to antibody directed against antigens on the cells. *Science* **169**, 68–70
7. Moolten, F.L., Capparell, N.J., Zajdel, S.H. & Cooperbrand, S.R. (1975) Antitumor effects of antibody-diphtheria conjugates II: immunotherapy with conjugates directed against tumor antigens

induced by Simian virus 40. *J. Natl. Cancer Inst.* **55**, 473–477

8.   Bacha, P., Murphy, J.R. & Reichlin, S. (1983) Thyrotropin-releasing hormone-diphtheria toxin-related polypeptide conjugates. *J. Biol. Chem.* **258**, 1565–1570

9.   Frankel, A.E. (ed.) (1988) *Immunotoxins*. Kluwer Academic Publishers, Dortrecht, Boston and Lancaster.

10. Vitetta, E.S., Thorpe, P.E. & Uhr, J.W. (1993) Immunotoxins: magic bullets or misguided missiles? *Immunol. Today* **14**, 252–259

11. Till, M.A., Ghetie, V., Gregory, T. *et al.* (1988) HIV-infected cells are killed by rCD4-ricin A chain. *Science* **242**, 1166–1168

12. Chaudhary, V.K., Mizukami, T., Feurst, T.R. *et al.* (1988) Selective killing of HIV-infected cells by recombinant human CD4-*Pseudomonas* exotoxin hybrid protein *Nature (London)* **335**, 369–372

13. Colombatti, M., Greenfield, L. & Youle, R.J. (1986) Cloned fragment of diphtheria toxin linked to T-cell specific antibody identifies regions of B chain active in cell entry. *J. Biol. Chem.* **261**, 3030–3035

14. Murphy, J.R., Bishai, W., Borowski, M., Miyanohara, A., Boyd, J. & Nagle, S. (1986) Genetic construction, expression and melanoma-sensitive cytotoxicity of a diphtheria toxin-related α-melanocyte stimulating hormone fusion protein. *Proc. Natl. Acad. Sci. U.S.A.* **83**, 8258–8261

15. Williams, D.P., Parker, K., Bacha, P. *et al.* (1987) Diphtheria toxin receptor binding domain substitution with interleukin 2: genetic construction and properties of a diphtheria toxin-related interleukin 2 fusion protein. *Protein Eng.* **1**, 493–498

16. LeMaistre, C.F., Meneghetti, C., Rosenblum, M. *et al.* (1992) Phase I trial of an interleukin-2 (IL-2) fusion toxin (DAB$_{486}$IL-2) in hematologic malignancies expressing the Il-2 receptor. *Blood* **79**, 2547–2554

17. Lakkis, F., Steele, A., Pacheco-Silva, A., Rubin-Kelley, V., Strom, T.B. & Murphy, J.R. (1991) Interleukin 4 receptor targeted cytotoxicity: genetic construction and *in vivo* immunosuppressive activity of a diphtheria toxin-related murine interleukin 4 fusion protein. *Eur. J. Immunol.* **21**, 2253–2258

18. Jean, L.L. & Murphy, J.R. (1991) Diphtheria toxin receptor-binding domain substitution with interleukin 6: genetic construction and interleukin 6 receptor-specific action of a diphtheria toxin-related interleukin 6 fusion protein. *Prot. Eng.* **4**, 989–994

19. Aullo, P., Alcami, J., Popoff, M.R., Klatzmann, D.R., Murphy, J.R. & Boquet, P. (1992) A recombinant diphtheria toxin related human CD4 fusion protein specifically kills HIV infected cells which express gp120 but selects fusion toxin resistant cells which carry HIV. *EMBO J.* **11**, 575–583

20. Murphy, J.R. & Strom, T.B. (1990) Diphtheria toxin-peptide hormone fusion proteins: protein engineering and selective action of a new class of recombinant biological response modifiers. In *ADP-ribosylating toxins and G proteins: Insights into signal transduction* (Moss, J. & Vaughan, M., eds.), pp. 141–160, Am. Soc. Microbiol., Washington

21. vanderSpek, J., Cosenza, L., Woodworth, T., Nichols, J.C. & Murphy, J.R. (1994) Diphtheria toxin-related cytokine fusion proteins: elongation factor 2 as a target for the treatment of neoplastic disease. *Mol. Cell. Biochem.* **138**, 151–156

22. Shaw, J.P., Akiyoshi, D.E., Arrigo, D.A. *et al.* (1991) Cytotoxic properties of DAB$_{486}$EGF and DAB$_{389}$EGF, epidermal growth factor (EGF) receptor-targeted fusion toxins. *J. Biol. Chem.* **266**, 21118–21124

23. Goodwin, R.G., Friend, D., Ziegler, S.F. *et al.* (1990) Cloning of the human and murine Interleukin-7 receptors: demonstration of a soluble form and homology to a new receptor superfamily. *Cell* **60**, 941–951

9

Thyrotropin-releasing hormone: basis and potential for its therapeutic use

Julie A. Kelly

Department of Biochemistry, Trinity College, Dublin 2, Ireland

Background to the clinical use of TRH

Thyrotropin-releasing hormone (TRH) was the first hypothalamic regulatory hormone to be characterized. It is a tripeptide with the primary structure L-pyroglutamyl-L-histidyl-L-proline amide (Glp-His-ProNH$_2$) (see Figure 1). Subsequently, TRH was shown to stimulate the release of thyroid-stimulating hormone (TSH) from the anterior pituitary, and to play a central role in regulating the pituitary–thyroid axis. It has also been reported to influence the release of other hormones, such as prolactin, vasopressin, insulin, noradrenaline, and adrenaline (Figure 2)[1].

Diagnostically, intravenous (i.v.) administration of TRH has been used to assess the integrity of the hypothalamic–pituitary–thyroid axis in clinical conditions associated with disturbances in thyroid hormone levels. However, with the recent introduction of sensitive assays for TSH the need for the TRH test has declined, except in cases where the capacity of the pituitary to secrete TSH is being examined[1,2].

Interestingly, the potential therapeutic applications of TRH that have attracted the most attention are not based on its endocrine properties, but on its broad spectrum of stimulatory actions within the central nervous system (CNS) (Table 1). These CNS-mediated effects provide the rationale for the use of TRH in the treatment of brain and spinal injury and certain CNS disorders, including Alzheimer's disease and motor neuron disease (MND)[2]. The hormonal actions of TRH, such as stimulation of the pituitary–thyroid axis

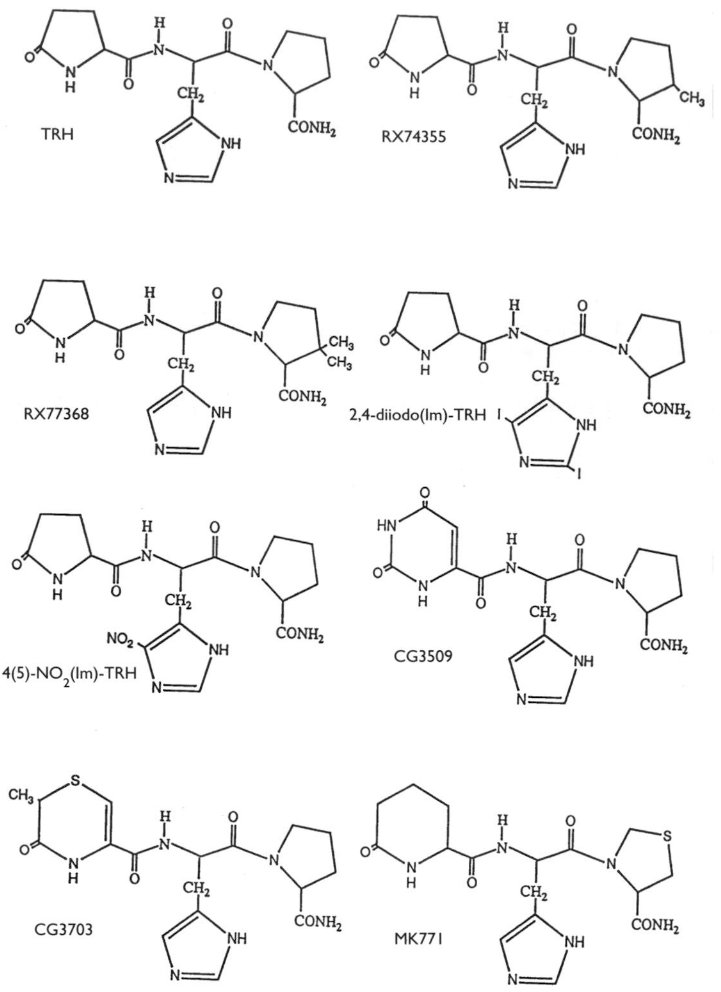

Figure 1. Some analogues of TRH

resulting in hyperthyroidism, are regarded as adverse side-effects in the treatment of CNS-related conditions.

The beneficial effects of TRH on CNS disorders and trauma appear to be due partly to its ability to potentiate other neurotransmitter systems and to reverse or attenuate certain actions of secondary injury factors that occur as a result of CNS trauma. However, the exact mechanism by which TRH improves these conditions is still not fully elucidated. It is clear though that these CNS effects of TRH are independent of the hypothalamic–pituitary–thyroid axis.

TRH has been found to be distributed widely in the CNS. Further, in the 1970s the involvement of TRH in CNS functions prompted a search for extra-pituitary TRH receptors that might mediate the CNS effects. It is now recognized that receptors for TRH are located throughout the CNS as well as several peripheral organs, such as the gastrointestinal tract. There is also growing evidence to suggest that, in addition to its role as a hormone, TRH can act as a neurotransmitter or neuromodulator in the CNS[1,3]. In potentiating the activity of other neurotransmitter systems, TRH appears to be acting as a neuromodulator at a number of different synapses (Figure 3).

Figure 2. Endocrine actions of TRH
TRH is released from the hypothalamic neurons into the hypothalamo-hypophyseal portal vessels that carry blood between the hypothalamus and the anterior pituitary. The anterior pituitary contains several different types of endocrine cells. TRH receptors are located on two anterior pituitary cell types: thyrotropes (TSH-releasing cells) and mammotropes (prolactin-releasing cells). Binding of TRH to the receptors on these cells stimulates the release of TSH and prolactin. The posterior lobe of the pituitary is composed of extensions of two groups of neurons whose cell bodies origin in the hypothalamus. These neurons extend via the pituitary stalk to the posterior lobe and are responsible for the secretion of two hormones: vasopressin and oxytocin.

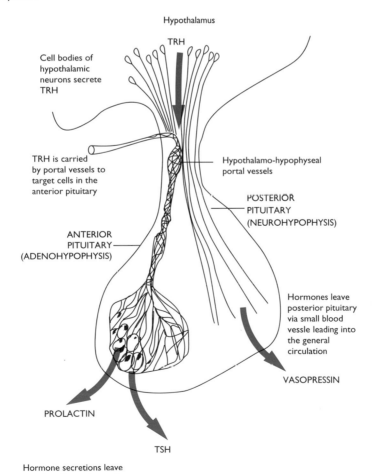

Table 1. Examples of the central stimulatory actions of TRH
•   Stimulation of locomotor activity
•   Increased blood pressure and heart rate
•   Increased cerebral blood flow
•   Increased respiratory rate
•   Increased gastric acid secretion and gastric emptying
•   Increased catecholamine release from the adrenal medulla
•   Stimulation of α-motor neuron activity
•   Stimulation of neuronal growth in culture

A case for the use of TRH in the treatment of CNS disorders and trauma

TRH was initially used to treat brain and spinal injury because of its ability to antagonize the damaging actions of endogenous opioids without reducing opioid analgesia. Opioids are believed to contribute to the secondary (or delayed) tissue damage often seen after CNS trauma[4,5]. The basis of the neuro-protective actions of TRH has now been extended to include its ability to antagonize the actions of other secondary injury factors, such as leukotrienes and platelet-activating factor, as well as to restore magnesium homoeostasis, improve bioenergetic status and decrease tissue oedema after brain injury. These beneficial effects and the therapeutic effectiveness of TRH in treating CNS trauma have been discussed by Faden and colleagues[4,5].

The mechanisms of the neuroprotective actions of TRH have yet to be elucidated fully. TRH has potent effects on blood flow that appear to be mediated through the CNS and it has been suggested that the antagonism of secondary injury factors may possibly be related to the beneficial effects of TRH on microcirculation[5].

Intracellular magnesium levels have been found to decline markedly after moderately severe brain trauma. Since magnesium ions are involved in the regulation of many cellular processes, a decrease in free intracellular magnesium could contribute to secondary tissue damage through a number of mechanisms, including impairment of energy metabolism, inhibition of Na^+/K^+-exchanging ATPase (EC 3.6.1.37) (which may contribute to cerebral oedema), calcium influx and perhaps modulation of opiate and excitatory amino acid receptor binding. It has been suggested that free magnesium may be chelated by fatty acids that are released during post-traumatic phospholipid hydrolysis, and that restoration of intracellular magnesium levels by TRH after traumatic brain injury may be brought about via binding to specific lipid metabolic products of phospholipase C (EC 3.1.4.3) activation[6].

Evidence that TRH may improve neurological outcome and survival after traumatic brain injury through direct effects on cerebral metabolism comes from studies in which the ratio of phosphocreatine to inorganic phosphate (PCr/P_i) was taken as an indirect measure of cellular energetic state. Using ^{31}P-

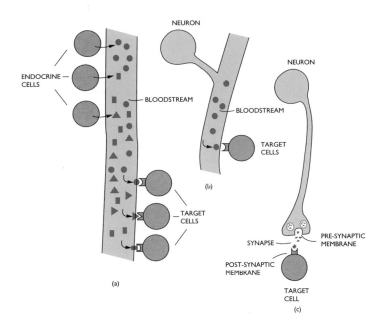

Figure 3. TRH as a neurohormone, neurotransmitter and neuromodulator
There are several modes of communication between cells, three of which are shown here. The first (a) is the least direct method of communication and involves the release of a chemical messenger, known in this case as a hormone, into the bloodstream which carries it to a target cell. Communication between the cell releasing the hormone, an endocrine cell, and the target cell is achieved by the binding of the hormone to receptors on the target cells which then elicit an appropriate response.

When a neuron or nerve cell releases a chemical messenger into the bloodstream the mode of communication is referred to as neuroendocrine communication and the messenger is known as a neurohormone (b). TRH acts as a neurohormone when it is released by hypothalamic neurons into the hypothalamo-hypophyseal portal vessels which carry it to target cells in the anterior pituitary.

The most direct mode of communication between cells is neurotransmission (c). In this case, the chemical messenger, referred to as a neurotransmitter, is released into the synaptic cleft and binds to specific receptors on the postsynaptic membrane of target cells. It has been suggested that TRH can act as a neurotransmitter in the CNS. In potentiating the activity of other neurotransmitter systems, TRH might also act as a neuromodulator in the CNS. It could achieve this in several ways, for example it might: (i) sensitize the postsynaptic receptors to the other (primary) neurotransmitter, (ii) increase the density of receptors for the primary neurotransmitter, or (iii) increase the release of the primary neurotransmitter from the presynaptic membrane. In such cases, the overall effect of the neuromodulator, TRH, would be to increase the effectiveness of neurotransmission by the primary neurotransmitter. It should be noted that a neuromodulator could also inhibit rather than facilitate neurotransmission by a primary neurotransmitter.

nuclear magnetic resonance spectroscopy, McIntosh and colleagues[7] have demonstrated that the PCr/P$_i$ ratio declines significantly after traumatic brain injury in rats. Treatment with an analogue of TRH increased the ratio in those animals showing enhanced neurological recovery, such that 4 h post-trauma the ratio was not significantly different from preinjury levels.

The use of TRH in the treatment of certain kinds of memory dysfunction, including Alzheimer's disease, is based on its ability to potentiate the activity of several neurotransmitter systems that are involved in memory, including cholinergic, noradrenergic, serotonergic and dopaminergic systems[3].

In an animal model for memory retrieval loss, Stwertka and colleagues[8] have shown intraperitoneal (i.p.) administration of TRH (0.1–30 mg/kg of body weight) protected against disruption of memory. They also demonstrated that the memory protection afforded by TRH was independent of a pituitary-mediated hormonal response. In a pilot study involving Alzheimer's disease patients, i.v. administration of TRH (0.3 mg/kg of body weight) significantly improved arousal and modestly improved semantic memory[9].

Furthermore, TRH has recently been shown to increase memory retention and retrieval in rats through activation of N-methyl-D-aspartate (NMDA)-receptor mediated processes[10]. NMDA-receptors belong to the group of receptors that mediate amino acid neurotransmission and appear to participate in long-term potentiation (LTP), a process that is believed to underlie learning and memory.

MND is characterized by progressive loss of voluntary motor function and is invariably fatal. Gradual degeneration of the large motor neurons in the cerebral cortex, brain stem and cervical and lumbar spinal cord occurs. Death usually results from respiratory failure due to bulbar or diaphragmatic involvement. Evidence showing that TRH has neurotrophic (neuron-growth promoting) and neuromodulatory effects in the motor neurones, together with its apparent beneficial effects in human spinocerebellar degenerations[3], would seem to indicate a possible role for TRH in the management of MND. Several groups, including Guiloff, Brooke, Brooks, Engel, and Munsat and colleagues initiated clinical trials to evaluate the effectiveness of TRH in the treatment of MND patients. Their results, presented at a New York Academy of Sciences conference on the Biomedical Significance of TRH, were initially seen as promising[11–15].

Guiloff found that i.v. administration of certain analogues of TRH resulted in slight improvements in weakness, spasticity, tongue movements, respiration and swallowing[11]. Whether a useful reduction in the rate of deterioration of the patients was achieved could not be determined by this study, though it appeared that the progression of the disease was not altered.

In reviewing the results from five different clinical studies, Brooke[12] observed that the design of no study was 'perfect'. Three of the studies showed transient, statistically significant effects in at least some muscles, whereas two studies using very small doses of TRH showed no such effects. He concluded that the effect of TRH in MND is a definite, acute, transient response requiring further investigation.

Brooks[13] also considered several clinical investigations and found both positive and negative results. In analysing the data he noted that the response of patients to TRH seemed to be influenced by whether they had bulbar or non-bulbar signs (assessment of bulbar functions include respiration, speech, swallowing and palatal, tongue and jaw movements), and also on whether they were male or female.

Engel discussed the reasons why clinical improvements with TRH or TRH analogues might be missed[14]. These included: (i) fixed-dose, fixed-schedule trials lacking flexibility for individualization; (ii) too low dosage regimes; (iii) too high doses/too infrequent administration; (iv) infrequent clinical testing at wrong timepoints; (v) the use of 'megascores', which are summated individual muscle scores that can obscure benefit in some muscle groups; and (vi) scoring of muscle strength between 1 and 5, which can miss slight but functionally useful improvement.

Engel also addressed the problem of 'autorefractoriness' which is associated with desensitization and down-regulation of the TRH receptor through repeated exposure to TRH and ultimately leads to unresponsiveness to TRH. Overall, he felt that TRH and TRH analogues can produce symptomatic benefit, such as increased muscle strength and decreased spasticity, for some patients with MND; however, he conceded that because of autorefractoriness TRH is currently difficult to harness for routine use[14].

Almost all the work of Munsat and colleagues has involved intrathecal administration of TRH in an attempt to slow the rate of deterioration in MND patients rather than producing short-term improvement in strength or function. The preliminary, uncontrolled observations that were presented at the conference indicated an apparent slowing of the rate of deterioration in the patients as well as short-lasting strength improvement[15]. However, more recent, comprehensive studies by this group show conflicting results[16]. These are discussed in more detail in the section entitled Intrathecal delivery systems (see later). Thus the value of TRH to produce long-term therapeutic benefits in MND patients remains unproven at this time.

There are a number of other clinical conditions for which TRH may have potential beneficial effects, including haemorrhagic, endotoxic and anaphylactic shock. In many cases the rationale for using TRH is the same as that outlined above, and is described in other reviews[2,3,17].

TRH as a neuropharmacological drug

As a peptide, several of the properties of TRH tend to undermine its therapeutic potential. Thus TRH is actually a poor candidate for use as a therapeutic agent in the treatment of CNS disorders and trauma.

In common with other peptide drugs, TRH is degraded very quickly by proteolytic enzymes. In brain and spinal cord, three enzymes catalyse the initial degradation of TRH[18] as shown in Figure 4. Two of these are soluble peptidases: pyroglutamyl aminopeptidase (PAP-I) (EC 3.4.19.3), which removes the N-terminal pyroglutamate from TRH; and prolyl oligopeptidase (EC 3.4.21.26), which deamidates TRH to give the free acid (acid TRH). When exposed to membrane fractions, TRH is degraded primarily by the ectoenzyme pyroglutamyl aminopeptidase II (PAP-II) (3.4.19.6). PAP-II has similar properties to thyroliberinase, the TRH-degrading enzyme in serum.

The physiological significance of the soluble enzymes is not clear, since neurotransmitter inactivation by enzymic degradation would probably occur outside of the neuron by ectoenzymes located on the cell surface, or within lysosomes after endocytosis. Ectoenzymes are thus strategically located to hydrolyse synaptically released peptides, and may represent a method of inactivation and regulation of peptide neurotransmitters such as TRH. The activity of the serum enzyme thyroliberinase appears to be controlled by thyroid hormones and may also be involved in the regulation of TRH[19]. It remains to be established whether the degradation of TRH during its transport in the hypothalamo-hypophyseal portal vessels represents a mechanism for controlling the amount of TRH reaching its target cells in the anterior pituitary.

Given the location of PAP-II on the synaptosomal and adenohypophyseal membranes, and its high degree of substrate specificity towards TRH, it seems likely that PAP-II may play a major role in inactivating TRH after its release[19]. O'Cuinn and colleagues have recently reviewed the actions and properties of the enzymes that degrade TRH[20].

The neuropharmacological effectiveness of TRH is also reduced by its relative lipid insolubility which hampers its passage across the blood–brain barrier. Limited permeability from bloodstream into brain is a property shared by TRH and other water-soluble peptides. The transport of peptides across the blood–brain barrier has been extensively considered by Banks and colleagues[21]. They observed that the degree of blood-to-brain passage for many peptides can be largely explained on the basis of their lipophilicity. However, recent studies demonstrate that the importance of peptide conformation in peptide–bilayer interactions may also be an important factor[22]. In spite of the restrictions to the passage of TRH across the blood–brain barrier, Zlokovic and colleagues have been able to show that tritiated TRH can be transported across the blood–brain barrier of guinea-pig brain by a non-saturable transport mechanism[23].

To overcome the instability of TRH and its low penetration across the blood–brain barrier, large doses can be required to achieve a neuropharmacological effect. In turn, these produce undesirable peripheral (including endocrine) effects. For example, side-effects observed during the treatment of MND patients with TRH include sweating, nausea, mild transient hypertension and a rise in thyroid hormones (although not to a level that the patients would be classified as clinically hyperthyroid)[15].

Several strategies are currently being investigated in an attempt to overcome these drawbacks. These include the use of the following: (i) TRH analogues; (ii) peptide mimetics; (iii) microspheres; (iv) intrathecal drug delivery systems; and (v) inhibitors of TRH-specific peptidases.

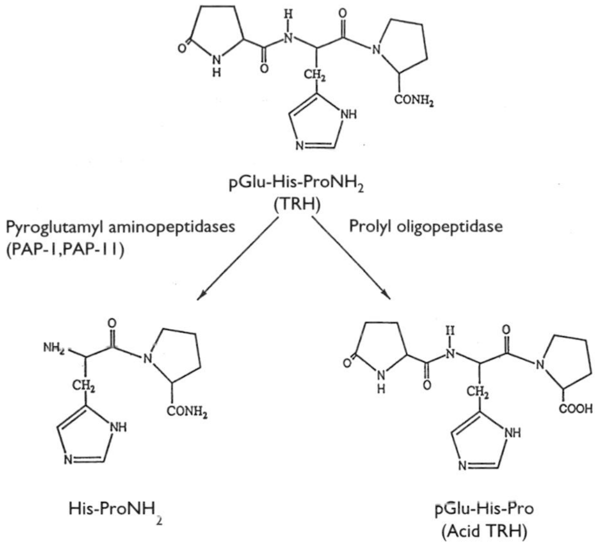

Figure 4. Metabolic pathway for the degradation of TRH in brain and spinal cord

Strategies to overcome limitations to the therapeutic use of TRH

TRH analogues

In an attempt to overcome the susceptibility of TRH to enzymic degradation, various analogues have been synthesized and evaluated. These have tended to incorporate modifications that confer stability to the pyroglutamyl amino-peptidases (PAPs) and/or prolyl oligopeptidase[24]. A selection of some of the available TRH analogues is shown in Figure 1. All of the analogues shown are agonist analogues of TRH that are capable of binding to and activating TRH receptors. In contrast, an antagonist analogue would bind to the receptor but would not activate it.

Studies *in vitro* show that methylation of prolineamide in TRH, as seen in RX77368 and RX74399, protects against deamidation by the prolyl oligopepti-dase. Alternatively, replacement of the pyroglutamyl residue with, for example, a six-membered ring as seen in the case of CG3509 and CG3703, confers resistance to the action of the PAPs[18,25].

Both the RX and CG analogues exhibit reduced affinity for central TRH receptors. However, structure–activity studies reveal that some of the analogues with a substituted pyroglutamyl moiety have enhanced CNS activity compared with TRH. This may be explained in part by their resistance to degradation by the PAPs. Conversely, the endocrinological activity of analogues in which the pyroglutamyl residue is replaced by a five- or six-membered ring is reduced compared with TRH[25]. These paradoxical results may also be partially explained by considering the ability of the analogues to assume certain three-dimensional conformations, which in turn influence their ability to bind favourably to the pituitary and/or CNS TRH receptors.

The ability of TRH and TRH analogues to bind to central or pituitary receptors is dependent on the three-dimensional conformation of the peptide and the interplay of a variety of forces, such as electrostatic interactions. Comparative conformation–activity relationships for hormonally and centrally acting TRH analogues now indicate that certain conformations are preferred at the CNS receptor, while others are preferred at the pituitary receptor (Figure 5)[26]. This is not to suggest that pituitary and CNS TRH receptors are, intrinsically, chemically different. It could be that other factors closely associated with TRH receptors in the membrane influence the binding of ligands to the receptor. It is certainly possible that, in common with other neurotransmitters, pharmacologically distinguishable types of TRH receptors exist.

Overall, these results indicate that it might be possible to design TRH analogues that bind selectively to either CNS or pituitary TRH receptors eliciting either a central or hormonal response, respectively. Stable, lipophilic agonist TRH analogues that are capable of selectively binding to central TRH receptors would be particularly useful for the treatment of CNS trauma and disorders.

Studies by Faden and colleagues[27] show that certain TRH analogues may improve the outcome after traumatic spinal cord injury. Further, the integrity of the C-terminal amino acid may be critical for the beneficial effects of TRH and TRH analogues. Thus CG3703 with a modified N-terminus significantly improves motor recovery after spinal trauma, whereas RX77368, which has a modified C-terminus, is without effect. This may reflect an ability of the CG3703 analogue to assume a three-dimensional conformation that would allow it to bind preferentially and to activate spinal cord TRH receptors. In addition, they have shown that two imidazole-substituted TRH analogues, $4(5)\text{-NO}_2(\text{Im})\text{TRH}$ and $2,4\text{-diiodo}(\text{Im})\text{TRH}$, improve behavioural recovery after experimental brain injury in rats. $2,4\text{-diiodo}(\text{Im})\text{TRH}$ was significantly more effective than $4(5)\text{-NO}_2(\text{Im})\text{TRH}$, which may be due to enhanced CNS penetration by the halogen-substituted analogue[28].

Peptide mimetics

Peptide mimetics are molecules that can mimic a natural peptide, such as TRH; however, because they incorporate non-peptide features in their structures, they tend to have increased stability, receptor selectivity and bioavailability compared with the endogenous peptide. They may either retain some features of the original peptide molecule or they may be completely non-peptide in structure.

Non-peptide ligands are only feasible if the peptide backbone is not essential for receptor binding. If this is shown to be the case, then information regarding structure–activity and conformation–activity relationships can be used to facilitate the rational design of peptide mimetics.

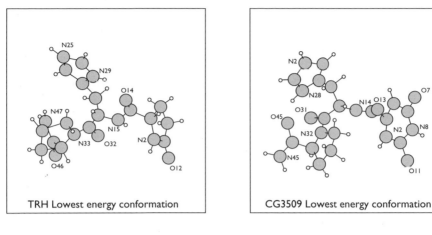

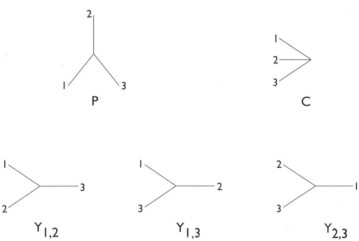

Figure 5. Conformations of TRH and TRH analogues

Computational studies[18] reveal that TRH conformers can be fitted into three classes depending on the relative position of the rings. The P conformer describes a propeller-like arrangement; the C conformer has the rings orientated like a cup; and the Y conformer has two rings in close proximity (the subscripts indicate which). 1 refers to the pyroglutamyl residue, 2 to the histidyl residue and 3 to the prolineamide residue.

The calculated lowest energy conformer for TRH has been found to be P. In contrast, that for CG 3509 has been found to be $Y_{2,3}$. The active form of TRH at the pituitary appeared to be the P conformer, whereas in the brain $Y_{2,3}$ was the most active conformer. CG 3509 was unable to assume the P conformer which correlates with its inability to stimulate the release of TSH from the pituitary.

Olson and colleagues[29] have developed peptide mimetics of TRH and TRH analogues in which the peptide backbone is replaced by a cyclohexane framework (Figure 6). These peptide mimetics, which have been shown to be active in cognitive performance, are devoid of endocrine activity and orally active. It remains to be established whether the cognitive enhancement activity

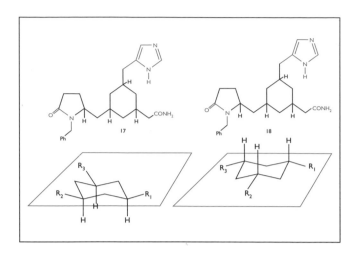

**Figure 6.
Orientations of the
side-chains in the
cyclohexane TRH
mimetic diastereo-
isomers synthe-
sized by Olson and
colleagues[29]**
Notice the pyro-
glutamylamide, pro-
lineamide and imida-
zole groups, which
are believed to be
important for TRH
activity, are present in
the peptide mimetic.

of these mimetics is achieved via binding to central TRH receptors or through another process. Nevertheless, it is encouraging that non-peptide compounds can be synthesized that mimic the biological functions of endogenous peptides. This approach will undoubtably lead to new, effective therapeutic agents.

Microspheres

From a practical standpoint, oral delivery of biopharmaceuticals is probably the most desirable mode of administration. However, this mode is not feasible for TRH as it is rapidly metabolized. Rather, it must be administered as a continuous high-dose i.v. infusion, or by daily injections.

In an attempt to reduce the frequency of administration, Heya and colleagues[30] have developed injectable microspheres, prepared with the biodegradable polymer copoly(D,L-lactic/glycolic acid) (PLGA) to produce a prolonged, controlled, injectable dosage form of TRH. They found that encapsulation of TRH in the microsphere protects it from enzymic degradation and allows TRH to be released into the blood over a long period of time.

One of the main difficulties associated with the release of water-soluble drugs, such as TRH, from microspheres is the large initial burst. Heya and colleagues overcame this by optimizing the concentration of TRH entrapped in the microsphere, the molecular mass of the PLGA and the lactic acid/glycolic acid ratio. Thus they were able to produce microspheres where, after a small initial burst, the release rate was dominated by the degradation of PLGA. Studies *in vivo* in rats showed that subcutaneous (s.c.) injection of these microspheres resulted in sustained release of TRH[30]. Overall, TRH-containing microspheres may well represent a convenient and reliable method for the delivery of TRH in the treatment of CNS disorders.

Intrathecal delivery systems

TRH and TRH analogues are usually administered by i.v. injection in the treatment of CNS disorders and trauma. Low concentrations of TRH are observed in the brain after this mode of administration, and these decrease rapidly. The influence of the route of administration on TRH brain concentrations has been investigated by Mitsuma and Nogimori[31]. They found that brain penetration rate was about 0.2% of the total dose administered following i.v., i.p., intramuscular (i.m.), or rectal administration. Higher brain TRH concentrations were achieved via rectal or i.m. administration, thus these routes of administration may be preferred for the therapeutic applications of TRH.

Munsat and colleagues explored the use of direct intrathecal administration in MND patients — suspecting that traditional routes of administration fail to achieve effective CNS concentrations and undermine the efficacy of TRH in MND treatment[16]. They were able to demonstrate that their intrathecal delivery system, an implanted constant-infusion pump, could deliver TRH into the cerebrospinal fluid at levels that had been shown to be neuropharmacologically effective in tissue culture experiments. This system proved to be safe, reliable and well-tolerated by patients. However, no alteration in the progressive course of MND was observed in the 36 patients studied over a period of 6 months of TRH treatment. As mentioned earlier, chronic exposure of the CNS to TRH can result in a rapid, reversible down-regulation of its receptors accompanied by the development of functional tolerance. Therefore, it is possible that TRH may be more effective as a trophic substance when given in pulses rather than as a constant infusion. Further studies are required to investigate this.

Inhibitors of TRH-specific enzymes

A major factor undermining the therapeutic use of TRH is its susceptibility to the action of certain peptidases in the CNS and serum. Inhibitors of TRH-degrading enzymes would protect from breakdown both endogenously released TRH or exogenously administered TRH or TRH analogues. Such inhibitors, therefore, might be therapeutically useful in potentiating the activity of endogenous TRH and exogenous TRH or TRH analogues.

The design of potent, selective inhibitors of a particular enzyme is facilitated by information regarding the substrate specificity and the mechanism of action of that enzyme. Proteases are divided into four classes according to their mechanism of action: (i) serine proteases, which have at their active centre histidine and serine residues involved in the catalytic process; (ii) cysteine proteases, which have a cysteine residue at the active site; (iii) aspartic proteases, which have a pH optimum below 5 due to the involvement of an acidic residue in the catalytic process; and (iv) metalloproteases, which use a metal ion in the catalytic mechanism.

Protease inhibitors often incorporate a structural feature that enables the inhibitor to interact with the active site of the enzyme. For example, a thiol group can interact with a zinc ion at the active site of a metalloprotease. A second structural feature resembling the substrate of the enzyme may also be incorporated into the inhibitor to facilitate the binding of the inhibitor with the substrate-binding site on the enzyme.

Perhaps one of the best examples of this approach is the design of active-site-directed inhibitors of angiotensin-converting enzyme (ACE)[32] (Figure 7). ACE catalyses the conversion of the decapeptide angiotensin I to angiotensin II, a powerful circulating vasoconstrictor. Inhibition of the production of angiotensin II has been found to be useful in the treatment of hypertension.

It is now recognized that TRH is primarily degraded by four enzymes: PAP-I, PAP-II, prolyl oligopeptidase and thyroliberinase. The specific mechanisms of action of these enzymes are not yet known. However, prolyl oligopeptidase appears to be a serine protease with a sensitivity to cysteine-blocking reagents, and PAP-I has been classed as a cysteine protease.

Bauer and colleagues have recently purified, characterized and cloned the cDNA of PAP-II, the ectoenzyme that exhibits a high degree of specificity to TRH[19,33]. From this work it can be seen that PAP-II can be classed as a zinc-metallopeptidase. Results such as these will greatly aid the rational design of site-directed PAP-II inhibitors and expand the understanding of the functional significance of this enzyme.

Although knowledge regarding the mechanism of action of the TRH-degrading enzymes is limited, the approach outlined above for ACE can be employed to design TRH-degrading enzyme inhibitors. Initial results are encouraging, and have recently been reviewed by Wilk[34]. Nevertheless, the development of TRH-degrading enzyme inhibitors is clearly in its infancy. More research needs to be carried out to produce potent and specific inhibitors and to evaluate their pharmacological properties.

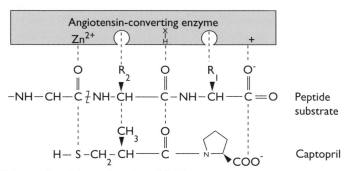

Figure 7. Schematic of the active site of ACE
The key features of the active site of ACE are shown together with the interactions between the substrate and the active site. This model was used to facilitate the design specific ACE inhibitors[24]. It can be seen that the potent ACE inhibitor captopril binds to the active site in a similar manner to the two terminal amino acids of the substrate. The terminal negatively-charged carboxyl group interacts with a positively charged residue on the enzyme, the amide carbonyl forms a hydrogen bond with a donor on the enzyme, and the proline ring and methyl side-chains interact with the enzyme in much the same way as the last two amino acid residues of the peptide substrate. However, the inhibitor interacts more strongly with the zinc ion of the enzyme because of its sulphydryl group.

Analogues that are stable to PAP activity have been shown to have enhanced CNS activity. This may indicate that PAP activity may be especially important in determining the biological half-life of TRH and TRH analogues. In contrast, alkylation of the amide nitrogen, which confers resistance to the prolyl oligopeptidase, does not result in analogues with prolonged CNS activity. This suggests that the prolyl oligopeptidase is of minor importance in the biological half-life of TRH and TRH analogues[25].

Conclusions and future direction

- *TRH has been shown to be much more than a hypothalamic regulatory hormone. More significantly, it has been shown to act as a neurotransmitter or neuromodulator in the CNS, to antagonize certain actions of secondary injury factors implicated in the delayed tissue damage after CNS trauma, and to exert trophic effects on spinal cord motor neurons. It is these properties that have established a basis for the use of TRH in the treatment of a number of CNS disorders and trauma.*
- *However, in common with other potential drugs that are peptides, the therapeutic efficacy of TRH is compromised by its instability and hydrophilic nature. Moreover, the high-dosage regimes needed to obtain a neuropharmacological effect can result in adverse side-effects arising from the endocrine and peripheral actions of TRH.*
- *Strategies are now being investigated to overcome these drawbacks and enhance the clinical effectiveness of TRH. Initial results are encouraging. Both peptide and non-peptide analogues of TRH are being designed and produced, and these appear to have increased stability, greater selectivity for central TRH receptors and enhanced lipophilicity compared with endogenous TRH.*
- *Improved TRH delivery methods have been developed, including an implanted, constant-infusion pump, intrathecal delivery system and TRH-containing microspheres. Further, inhibitors of TRH-degrading enzymes are being designed that may prove to be therapeutically useful in potentiating the activity of endogenous TRH and exogenously administered TRH or TRH analogues.*
- *Overall, information regarding the role of TRH in the CNS and the exact nature of its central actions is constantly expanding. This should facilitate advances in the strategies being explored to maximize the therapeutic effectiveness of TRH. Reciprocally, the development of TRH analogues and TRH-degrading enzyme inhibitors may aid the investigation of the interactions occurring at the TRH receptors and at the active site of the TRH-degrading enzymes. These may also be used to clarify the functional role of TRH-degrading enzymes.*

- *Although not discussed in this review, an approach that might be considered in future, as the regulation of TRH biosynthesis becomes better understood, is the manipulation of TRH production at the genome level.*

- *As the number of sites at which TRH and TRH-like peptides are identified in the body increases, so do the potential clinical applications for such peptides. For instance, TRH-related peptides have been located in the prostate complex and semen. Although their function in this region has not yet been elucidated, it appears that a role in reproduction is possible, with implications for their use in human infertility[35].*

- *In summary, the outlook for the therapeutic use of TRH and TRH analogues appears promising. Further, strategies now being employed to maximize the therapeutic efficacy of TRH will undoubtably be relevant to other peptides with potential clinical applications.*

References

1. Griffiths, E.C. (1985) Thyrotrophin releasing hormone: endocrine and central effects. *Psychoneuroendocrinol.* **10**, 225–235
2. Griffiths, E.C. (1987) Clinical applications of thyrotrophin-releasing hormone. *Clin. Sci.* **73**, 449–457
3. Horita, A., Carino, M.A. & Lai, H. (1986) Pharmacology of thyrotropin-releasing hormone. *Annu. Rev. Pharmacol. Toxicol.* **26**, 311–332
4. Faden, A.I. & Salzman, S. (1992) Pharmacological Strategies in CNS trauma. *Trends Pharmacol. Sci.* **13**, 29–35
5. Faden, A.I., Vink, R. & McIntosh, T.K. (1989) Thyrotropin-releasing hormone and central nervous system trauma. *Ann. N.Y. Acad. Sci.* **553**, 380–383
6. Vink, R., McIntosh, T.K. & Faden, A.I. (1988) Treatment with the thyrotropin-releasing hormone analog CG3703 restores magnesium homeostasis following traumatic brain injury in rats. *Brain Res.* **460**, 184–188
7. McIntosh, T.K., Vink, R. & Faden, A.I. (1988) An analogue of thyrotropin-releasing hormone improves outcome after brain injury: ^{31}P-NMR studies. *Am. J. Physiol.* **254**, R785–R792
8. Stwertka, S.A., Vincent, G.P., Gamzu, E.R., MacNeil, D.A. & Verderese, A.G. (1991) TRH protection against memory retrieval deficits is independent of endocrine effects. *Pharmacol. Biochem Behav.* **41**, 145–152
9. Mellow, A.M., Sunderland, T., Cohen, R.M., *et al.* (1989) Acute effects of high-dose thyrotropin releasing hormone infusions in Alzheimer's disease. *Psychopharmacology* **98**, 403–407
10. Kasparov, S.A. & Chizh, B.A. (1992) The NMDA-receptor antagonist Dizocilpine (MK-801) suppresses the memory facilitatory action of thyrotropin-releasing hormone. *Neuropeptides* **23**, 87–92
11. Guiloff, R.J. (1989) Use of TRH analogues in motor neuron disease. *Ann. N.Y. Acad. Sci.* **553**, 399–421
12. Brooke, M.H. (1989) Thyrotropin-releasing hormone in ALS: are the results of clinical studies inconsistent? *Ann. N.Y. Acad. Sci.* **553**, 422–430
13. Brooks, B.R. (1989) A summary of the current position of TRH in ALS therapy. *Ann. N.Y. Acad. Sci.* **553**, 431–461
14. Engel, K. (1989) High-dose TRH treatment of neuromuscular diseases: summary of mechanisms and critique of clinical studies. *Ann. N.Y. Acad. Sci.* **553**, 462–472

15. Munsat, T.L., Lechan, R., Taft, J.M., Jackson, M.D. & Reichlin, S. (1989) TRH and diseases of the motor system. *Ann. N.Y. Acad. Sci.* **553**, 388–398

16. Munsat, T.L., Taft, J., Jackson, I.M.D., *et al.* (1992) Intrathecal thyrotropin-releasing hormone does not alter the progressive course of ALS. *Neurology* **42**, 1049–1053

17. Holaday, J.W., Long, J.B., Martinez-Arizala, A., Chen. H.-S., Reynolds, D.G. & Gurll, N.J. (1989) Effects of TRH in circulatory shock and central nervous system ischemia. *Ann. N.Y. Acad. Sci.* **553**, 370–379

18. Griffiths, E.C., Kelly, J.A., Ashcroft, A., Ward, D.J. & Robson, B. (1989) Comparative metabolism and conformation of TRH and its analogues. *Ann. N.Y. Acad. Sci.* **553**, 217–231

19. Bauer, K. (1994) Purification and characterization of the thyrotropin-releasing-hormone-degrading-ectoenzyme. *Eur. J. Biochem.* **224**, 387–396

20. O'Cuinn, G., O'Connor, B. & Elmore, M. (1990) Degradation of thyrotropin-releasing hormone and luteinising hormone-releasing hormone by enzymes of brain tissue. *J. Neurochem.* **54**, 1–13

21. Banks, A., Kastin, A.J. & Barrera, C.M. (1991) Delivering peptides to the central nervous system: dilemmas and strategies. *Pharm. Res.* **8**, 1345–1350

22. Ramaswami, V., Haaseth, R.C., Matsunaga, T.O., Hruby, V.J. & O'Brien, D. (1992) Opioid peptide interactions with lipid bilayer membranes. *Biochim. Biophys. Acta* **1109**, 195–202

23. Zlokovic, B.V., Lipovac, M.N., Begley, D.J., Davson, H. & Rakic, L. (1988) Slow penetration of thyrotropin-releasing hormone across the blood–brain barrier of an *in situ* perfused guinea pig brain. *J. Neurochem.* **51**, 252–257

24. Metcalf, G. (1982) Regulatory peptides as a source of new drugs — the clinical prospects for analogues of TRH which are resistant to metabolic degradation. *Brain Res. Rev.* **4**, 389–408

25. Flohe, L., Bauer, K., Friderichs, E., *et al.* (1983) Biological effects of degradation-stabilised TRH analogues. In *Thyrotropin-Releasing Hormone* (Griffiths, E.C. & Bennnett, G., eds.), pp. 327–430, Raven Press, New York

26. Ward, D.J., Finn, P.W., Griffiths, E.C. & Robson, B. (1987) Comparative conformation–activity relationships for hormonally- and centrally-acting TRH analogues. *Int. J. Pept. Protein Res.* **30**, 263–274

27. Faden, A., Sackson, I. & Noble, L.J. (1988) Structure–activity relationships of TRH analogs in rat spinal cord injury. *Brain Res.* **448**, 287–293

28. Faden, A., Labroo, V.M. & Cohen, L.A. (1993) Imidazole-substituted analogues of TRH limit behavioral deficits after experimental brain trauma. *J. Neurotrauma* **10**, 101–107

29. Olson, G.L., Bolin, D.R., Bonner, M.P., et al. (1993) Concepts and progress in the development of peptide mimetics. *J. Med. Chem.* **36**, 3039–3049

30. Heya, T., Okado, H., Yasuaki, O. & Toguchi, H. (1991) Factors influencing the profiles of TRH release from copoly(D,L-lactic/glycolic acid) microspheres. *Int. J. Pharm.* **72**, 199–205

31. Mitsuma, T. & Nogimori, T. (1993) Influence of the route of administration on thyrotropin-releasing hormone concentration in the mouse brain. *Experientia* **39**, 620–622

32. Cushman, D.W., Cheung, H.S., Sabo, E.F. & Ondetti, M.A. (1977) Design of potent competitive inhibitors of angiotensin-converting enzyme: carboxyalkanoyl and mercaptoalkanoyl amino acids. *Biochemistry* **16**, 5484–5491

33. Schauder, B., Schomburg, L., Kohrle, J. & Bauer, K. (1994) Cloning of a cDNA encoding an ectoenzyme that degrades thyrotropin-releasing hormone. *Proc. Natl. Acad. Sci. U.S.A.* **91**, 95434–9538

34. Wilk, S. (1989) Inhibitors of TRH-degrading enzymes. *Ann. N.Y. Acad. Sci.* **553**, 252–264

35. Thetford, C.R., Morrell, J.M. & Cockle, S. (1992) TRH-related peptides in the rabbit prostate complex during development. *Biochim. Biophys. Acta* **1115**, 252–258

Subject index

A

N-Acetylglucosamine, 66, 69
Actin, 82
Acute lymphoblastic leukaemia, 126
Acute myelogenous leukaemia, 126
Adenosine, 27
Adhesion, of cell to target organ, 69
ADP/ATP translocator, 2
ADP-ribosyl cyclase, 107
ADP-ribosylation, 121–123
ADP-ribosyltransferase, 122
Adrenal chromaffin cell, 85
Adrenaline, 15
Affinity chromatography, 111
ALL (see Acute lymphoblastic
    leukaemia)
Alternative mRNA splicing, 23, 109,
    110
Alternative transcription initiation, 109
Alzheimer's disease, 133, 138
AML (see Acute myelogenous
    leukaemia)
Anaesthetic, 101
Angiotensin, 27, 146
Animal lectin, 60
Annexin, 88
Anti-adhesion drug, 63
Anti-inflammatory drug, 73
Antisense oligonucleotides, 89
Antizyme, 39–41
Antizyme inhibitor, 39
Aromatic L-amino acid decarboxylase,
    16

B

Bacterial lectin, 60

Bacterial toxin, 60, 120
Benzodiazepine receptor, 2, 10
Bilayer, 106
Bipolar disorder, 31, 113
Blood–brain barrier, 140
Botulinum neurotoxin, 87, 90
Bradykinin, 80

C

C-type lectin, 67, 69
Ca/CaMPKII (see Ca^{2+}/calmodulin-
    dependent protein kinase)
Ca^{2+}/calmodulin-dependent protein
    kinase, 26–28, 112
cADPR (see Cyclic ADP-ribose)
Caffeine, 101, 108
Calcium
    channel, 83, 97–114
    cytosolic concentration, 81, 97
    imaging, 81
    -induced Ca^{2+} release, 113
    probe, 98
    role in lectin binding, 65
cAMP-dependent kinase, 108
Cancer, 68, 73
Capacitative coupling, 106
Captopril, 146
4a-Carbinolamine tetrahydrobiopterin,
    17
Carbohydrate-recognition domain, 67,
    69
CAT (see Chloramphenicol
    acetyltransferase)
Catalytic domain
    of protein toxin, 120
    of tyrosine hydroxylase, 25

V

Vascular smooth muscle, 107
Vasoactive intestinal polypeptide, 27
VDAC (see Voltage-dependent anion
     channel)
Vestibule, of ion channel, 105
VIP (see Vasoactive intestinal
     polypeptide)
Voltage-clamped bilayer, 100
Voltage-dependent anion channel, 2
Voltage-sensitive cation channel, 112

X

X-ray crystallography, 61
Xenopus oocyte, 23

Essays in Biochemistry: Volume 29

Edited by **D K Apps**, *University of Edinburgh* and **K F Tipton**, *Trinity College, Dublin*

Studying for a either a module or a degree in Biochemistry? Want to up-date your lecture notes with the latest research findings?

Essays in Biochemistry will provide you with a single source of information on the latest research in rapidly moving areas of biochemistry and molecular biology. Particular attention is given to topics which perhaps because of recent developments are poorly covered by standard texts. The extensive bibliographies and further reading lists will provide undergraduates and postgraduates students with a valuable bridge between text-books and research papers.

Each chapter is written by experts in their area of research and is a self-contained summary of the state-of-the-art of that topic.

Illustrated throughout with clear diagrams this highly readable volume will also serve as a valuable teaching aid to lecturers in biochemistry and molecular biology.

Contents: Bacterial DD-transpeptidases and penicillin, M Jamin, J-M Wilkin and J-M Frère; Sphingolipid activator proteins, K Suzuki; Oxygen toxicity, free radicals and antioxidants in human disease: biochemical implications in atherosclerosis and the problems of premature neonates, C A Rice-Evans and V Gopinathan; Reconstructed human skin: transplant, graft or biological dressing? E J Wood and I R Harris; Opsin genes, B E H Maden; The roles of molecular chaperones *in vivo*, P A Lund; Molecular chaperones: physical and mechanistic properties, S G Burston and A R Clarke; Affinity precipitation: a novel approach to protein purification, J A Irwin and K F Tipton; Molecular pathology of prion diseases, C Smith and J Collinge; Ribozymes, H A James and P C Turner; Protein stability at high temperatures, D A Cowan; Subject index.

ISBN 1 85578 017 8   Paperback   230 pages   £17.50/US$30.00   May 1995

Essays in Biochemistry: Volume 28

Edited by **K F Tipton**, *Trinity College, Dublin*

Contents: Metabolic control, M Slater, R G Knowles and C I Pogson; The role of mitochondrial HMG-CoA synthase in regulation of ketogenesis, P A Quant; Motor neurone disease, A Goonetilleke, J de Belleroche and R J Guiloff; Carnitine and its role in acyl group metabolism, R R Ramsay; Folate/vitamin B_{12} inter-relationships, J Scott and D Weir; Protein kinase inhibitors, H Hidaka and R Kobayashi; Mitochondrial DNA and disease, S R Hammans; PIG-tailed membrane proteins, A J Turner; Horseradish peroxidase: the analyst's friend, O Ryan, M R Smyth and C Ó Fágáin; The renin – angiotensin system, T Inagami; Subject index.

ISBN 1 85578 016 X   Paperback   185 pages   £17.50/US$30.00   May 1994

Orders and enquiries to: **Portland Press Ltd**, Commerce Way, Colchester CO2 8HP, UK
Tel: (01206) 796351 Fax: (01206) 799331
In North America : **Portland Press Inc**, Ashgate Publishing Co, Old Post Road, Brookfield, VT 05036-9704, USA   Fax (802) 276 3837 Tel (802) 276 3162
* Carriage: UK customers please add £2.00 per order, overseas customers add £3.00, US customers add US$3.50 per book

Fundamentals of Enzyme Kinetics

A Cornish-Bowden, *LCB-CNRS, Marseilles, France*

Since the first edition was published in 1979, the development of techniques for studying and manipulating genes has transformed biochemistry. Nonetheless, enzymes remain at the heart of all living systems, and an understanding of how they operate is vital for understanding the chemistry of life.

This book describes the principles of enzyme kinetics, with an emphasis on the fundamentals, rather than an encyclopaedic accumulation of facts, to allow readers to fill in gaps for themselves and proceed in the subject as far as they need to. In this way it provides the basis for understanding enzyme kinetics, whether at the level of the undergraduate, the research student or the researcher.

Contents: Basic principles of chemical kinetics; Introduction to enzyme kinetics; Practical aspects of kinetic studies; How to derive steady-state rate equations; Inhibition and activation of enzymes; Reactions of more than one substrate; Use of isotopes for studying enzyme mechanisms; Environmental effects on enzymes; Control of enzyme activity; Kinetics of multi-enzyme systems; Fast reactions; Estimation of kinetic constants; References; Solutions to problems; Index.

1 85578 072 0   Paperback   July 1995   352 pages   £18.00/US$29.00

Postgraduate Study in the Biological Sciences

A Researcher's Companion

By **R J Beynon,** *UMIST, Manchester*

This book will excel as a foundation course on essential skills for research students that, although seen as peripheral to experimentation, mark a professional scientist. Aimed at both student and supervisor, it contains basic but well-founded advice and information that will be needed by newcomers to experimental science. Used as a supplement to established practices, it aims to extend student awareness and skills though their own efforts.

" *I have encountered no other book quite like this one for brevity, readability and practical down-to earth good sense* "   **Times Higher Education Supplement**

"*... sound advice on every aspect of doing a higher degree*"
    Trends in Cell Biology

Contents: Starting out. Developing your experimental skills. Conducting a research project. Writing about your work. Talking about your work. Computers and computing. Safety matters. The graduate student as teacher. Moving on.

1 85578 009 7   Paperback   May 1993   150 pages   £9.50/US$17.00

Orders and enquiries to: **Portland Press Ltd**, Commerce Way, Colchester CO2 8HP, UK
Tel: (01206) 796351 Fax: (01206) 799331
In North America : **Portland Press Inc**, Ashgate Publishing Co, Old Post Road, Brookfield, VT 05036-9704, USA   Fax (802) 276 3837 Tel (802) 276 3162
* Carriage: UK customers please add £2.00 per order, overseas customers add £3.00, US customers add US$3.50 per book